T0136850

Springer Theses

Recognizing Outstanding Ph.D. Research

Aims and Scope

The series "Springer Theses" brings together a selection of the very best Ph.D. theses from around the world and across the physical sciences. Nominated and endorsed by two recognized specialists, each published volume has been selected for its scientific excellence and the high impact of its contents for the pertinent field of research. For greater accessibility to non-specialists, the published versions include an extended introduction, as well as a foreword by the student's supervisor explaining the special relevance of the work for the field. As a whole, the series will provide a valuable resource both for newcomers to the research fields described, and for other scientists seeking detailed background information on special questions. Finally, it provides an accredited documentation of the valuable contributions made by today's younger generation of scientists.

Theses are accepted into the series by invited nomination only and must fulfill all of the following criteria

- They must be written in good English.
- The topic should fall within the confines of Chemistry, Physics, Earth Sciences, Engineering and related interdisciplinary fields such as Materials, Nanoscience, Chemical Engineering, Complex Systems and Biophysics.
- The work reported in the thesis must represent a significant scientific advance.
- If the thesis includes previously published material, permission to reproduce this must be gained from the respective copyright holder.
- They must have been examined and passed during the 12 months prior to nomination.
- Each thesis should include a foreword by the supervisor outlining the significance of its content.
- The theses should have a clearly defined structure including an introduction accessible to scientists not expert in that particular field.

More information about this series at http://www.springer.com/series/8790

Chiara Rizzi

Searches for Supersymmetric Particles in Final States with Multiple Top and Bottom Quarks with the Atlas Detector

Doctoral Thesis accepted by
Universitat Autònoma de Barcelona,
Barcelona, Spain

 Springer

Author
Dr. Chiara Rizzi
Experimental Physics Department
CERN
Geneva, Switzerland

Departament de Física, Institut de Física
d'Altes Energies, Facultat de Ciències
Universitat Autònoma de Barcelona
Bellaterra, Barcelona, Spain

Supervisor
Prof. Aurelio Juste Rozas
ICREA/Institut de Física d'Altes Energies
Universitat Autònoma de Barcelona
Barcelona, Spain

ISSN 2190-5053 ISSN 2190-5061 (electronic)
Springer Theses
ISBN 978-3-030-52879-9 ISBN 978-3-030-52877-5 (eBook)
https://doi.org/10.1007/978-3-030-52877-5

This Springer imprint is published by the registered company Springer Nature Switzerland AG
The registered company address is: Gewerbestrasse 11, 6330 Cham, Switzerland

Supervisor's Foreword

This Ph.D. thesis documents two of the highest-profile searches for supersymmetry performed at the ATLAS experiment using up to 80 fb^{-1} of proton-proton collision data at a center-of-mass energy of 13 TeV delivered by the Large Hadron Collider (LHC) during its Run 2 (2015–2018). The signals of interest feature a high multiplicity of jets originating from the hadronisation of b-quarks and large missing transverse momentum, which constitutes one of the most promising final state signatures for discovery of new phenomena at the LHC. The first search is focused on the strong production of a pair of gluinos, with each gluino decaying into a neutralino and a top-antitop-quark pair or a bottom-antibottom-quark pair. The second search targets the pair production of higgsinos, with each higgsino decaying into a gravitino and a Higgs boson, which in turn is required to decay into a bottom-antibottom-quark pair. Both searches employ state-of-the-art experimental techniques and analysis strategies at the LHC, resulting in some of the most restrictive bounds available to date on the masses of the gluino, neutralino, and higgsino in the context of the models explored.

Barcelona, Spain   Prof. Aurelio Juste Rozas

Abstract

This dissertation presents two searches for Supersymmetry in proton-proton collisions at CERN's LHC, targeting signal models that lead to the production of multiple top quarks or bottom quarks in the final state.

The first search targets gluino pair production, where each gluino decays through a top squark (Gtt model) or a bottom squark (Gbb model) to a top-antitop quark pair or a bottom-antibottom quark pair, respectively, and a neutralino, which is the Lightest Supersymmetric Partner (LSP). Each top quark in turn decays to a W boson and an a bottom quark. Thus, the final state is characterized by a high multiplicity of b-jets, which are collimated sprays of particles originating from the hadronization of bottom quarks, and missing transverse momentum from the LSP that escapes the detection.

The second search targets a GGM model of higgsino pair production, where each higgsino decays to a Standard Model Higgs boson and a gravitino, which in this case plays the role of the LSP. This search focuses on the decay of the two Higgs bosons to bottom-antibottom quark pairs, yielding again final states with multiple b-jets. This is the first ATLAS analysis targeting this signature, which had been previously considered in searches performed by the CMS Collaboration.

Both searches in the thesis use the data collected by the ATLAS experiment at the LHC between 2015 and 2016, at a center-of-mass energy $\sqrt{s} = 13$ TeV, corresponding to an integrated luminosity of 36.1 fb^{-1}. The gluino search, without further reoptimization, is also extended using the data collected in 2017, for a total integrated luminosity of 79.8 fb^{-1}.

No significant excess of events above the Standard Model expectation is observed in any of the search regions, and the results are used to set upper limits on the production of supersymmetric particles. The first search excludes at 95% confidence level gluino masses up to 2.25 TeV for the Gtt model and up to 2.17 TeV for the Gbb model, in both cases for neutralino masses below 800 GeV. The second search excludes higgsino masses in the range 240–880 GeV, assuming that the higgsino decays exclusively to a Higgs boson and a gravitino.

Acknowledgements

During the last few years, I have had the opportunity to work and share fun times with a lot of brilliant and kind people, and this has really been one of the best aspects of my Ph.D.

First of all, I would like to thank my Ph.D. supervisor, Aurelio Juste, who has always been extremely supportive of my work; he has let me free to explore my ideas, while being constantly available to provide guidance and advice. Aurelio and Martine Bosman have been able to create a great working environment, and all the people who are (or have been) part of the IFAE group have been precious colleagues and good friends. All the post-docs in the group have always been available to help me with anything from technicalities to long physics discussions, and I am particularly indebted to Loïc Valery and Trisha Farooque, who have helped me in growing as a scientist and have offered me their friendship. Carlos, Davide, Julian, Mirko, Nicola and Tal have been great office mates and a pleasure to spend the time with.

This Ph.D. has not been the beginning of my experience in ATLAS, and I am extremely grateful to Tommaso Lari, who supervised me during my bachelor and master thesis at the University of Milan, and continued being a reference figure also during my Ph.D.; he taught me how to appreciate the most interesting aspects of High Energy Physics research and how to have fun while doing it.

Since the beginning of the Ph.D., I have found in the multi-b team a group of people who kept me motivated and helped me in gaining independence in my work, in an atmosphere that always enhanced collaboration. I am particularly grateful to Max Swiatlowski, whose expertise, energy and dedication have been invaluable ingredients in building a wonderful team, and to Giordon Stark, who has always been available to kindly help me and many others (and to whom goes the credit for the awesome event display on the cover).

I have had the chance to work with many great people in the ATLAS SUSY group, and I want to thank particularly Iacopo Vivarelli, Till Eifert and Zach Marshall, who have been conveners of the group, and all the subconveners who in the past years have put a lot of effort and passion in organizing the group and providing support and critical review to each analysis.

I would like to thank the people working in TileCal luminosity group, and I am especially grateful to Ilya Korolkov, Richard Hawkings and Witold Kozanecki for their patience and their inspiring ideas. I also would like to thank all the people of the HistFitter team, particularly Sophio Pataraia and Jeanette Lorenz.

Moving to a new place is always somehow scary, but since the beginning of my stay at CERN in 2014, I have had the opportunity to rely on Andrea, a fellow Ph.D. student in the IFAE group. Her amazing human skills and optimism have helped me making a lot of good friends, including Dott. Ing. Carlo Zanoni, who has been an important part of my everyday life.

During the last few years, a lot of people that are currently studying in Milan, or that like me studied in Milan to then continue working somewhere else, created a lively CERN-based community that makes me feel closer to home. Thanks to Alessandra, Carlo, Davide, Ettore, Giacomo, Giulia, Maria Giulia, Stefano and the rest of the lunch crew! A special thank you goes to the SUSY-enthusiast Claudia, Sonia and Lorenzo who have helped me a lot in the last-minute panic moments.

My family has been supportive of my choices for my entire life. Without the help and love of my mom, my dad and Laura, I would not have even been able to start this work. Thank you for giving me all the opportunities I could ask for.

Throughout these years, I have had one person who never doubted that I would make it, who lived with me through the darkest times of missed deadlines and who still makes me smile every single day independently of everything else. Thank you Giuseppe for knowing me better than anybody else, and deciding to share your life with me!

Contents

Acronyms

AMSB	Anomaly-mediated supersymmetry breaking
ATLAS	A Toroidal LHC Apparatus
BCID	Bunch-crossing identifier
BCM	Beam Conditions Monitor
BDT	Boosted decision tree
BR	Branching ratio
BSM	Beyond Standard Model
CERN	European Organization for Nuclear Research
CKM	Cabibbo–Kobayashi–Maskawa
CL	Confidence level
CL_s	Modified confidence level
CMS	Compact Muon Solenoid
corrJVF	corrected jet vertex fraction
CR	Control region
CSC	Cathode strip chambers
CST	Calorimeter soft term
CTP	Central Trigger Processor
DR	Diagram removal
DS	Diagram subtraction
EB	Extended barrel
ECal	Electromagnetic calorimeter
EM	Electromagnetic
FSR	Final-state radiation
GGM	General gauge mediation
GMSB	Gauge-mediated supersymmetry breaking
GRL	Good run list
GSC	Global sequential calibration
HCal	Hadronic Calorimeter
HLT	High-level trigger
HPD	Hybrid photon detectors

IBL	Insertable B-Layer
ID	Inner Detector
IP	Interaction point
IR	Infrared
ISR	Initial-state radiation
JER	Jet energy resolution
JES	Jet energy scale
JVF	Jet vertex fraction
JVT	Jet vertex tagger
L1	Level 1
L1Calo	Level 1 Calorimeter trigger
L1Muon	Level 1 Muon trigger
L1Topo	Level 1 Topological trigger
LAr	Liquid Argon
LB	Long barrel
LB	Luminosity block
LEIR	Low Energy Ion Ring
LEP	Large Electron-Positron Collider
LHC	Large Hadron Collider
LL	Leading logarithm
LLR	Log-likelihood ratio
LO	Leading order
LSP	Lightest supersymmetric partner
MC	Monte Carlo
MCP	Microchannel plates
MDT	Monitored Drift Tubes
ME	Matrix element
MLE	Maximum likelihood estimate
MS	Muon spectrometer
MSSM	Minimal Supersymmetric Standard Model
NLL	Next-to-leading logarithmic
NLO	Next-to-leading order
NLSP	Next to lightest supersymmetric partner
NNLL	Next-to-next-to-leading logarithmic
NNLO	Next-to-next-to-leading order
NP	Nuisance parameters
OP	Operating point
OR	Overlap removal
PDF	Parton distribution function
PDF	Probability density function
PLR	Profiled likelihood ratio
PMSB	Planck-scale-mediated supersymmetry breaking
pMSSM	phenomenological Minimal Supersymmetric Standard Model
PMT	Photonmultiplier tube
POI	Parameter of interest

pp	Proton-proton
PS	Parton shower
PS	Proton Synchrotron
PSB	Proton Synchrotron Booster
PV	Primary vertex
QCD	Quantum chromodynamics
QED	Quantum electrodynamics
RDO	Raw data object
RF	Radiofrequency
RMS	Root mean square
RoI	Region of interest
RPC	R-parity conserving
RPC	Resistive plate chambers
RPV	R-parity violating
SCT	Semiconductor Tracker
SF	Scale factor
SM	Standard Model
SPS	Super Proton Synchrotron
SR	Signal region
SSB	Spontaneous symmetry breaking
SUSY	Supersymmetry
TDAQ	Trigger and data acquisition
TF	Transfer factor
TGC	Thin gap chambers
TileCal	Tile barrel calorimeter
TRD	Transition radiation detector
TRT	Transition Radiation Tracker
TST	Track soft term
UL	Upper limit
vdM	van der Meer
VEV	Vacuum expectation value
VR	Validation region

Chapter 1
Preamble

At the Large Hadron Collider (LHC) proton-proton (*pp*) collisions are used to probe the nature of particles in energy regimes that were not accessible before. After the discovery of the Higgs boson in 2012, the LHC is now continuing its operations with the dual goal of measuring the Higgs boson properties in great detail but also to continue the quest for new particles.

ATLAS is one of the two general purpose experiments on the LHC ring, and this thesis focuses on the data that is has collected between 2015 and 2016, at a center-of-mass energy $\sqrt{s} = 13$ TeV, corresponding to an integrated luminosity of approximately 36 fb^{-1}.

The Standard Model (SM) is the theory that as of today best describes the experimental results on subatomic particles. Nevertheless, there are strong theoretical and experimental arguments to believe that the SM is the low-energy limit of a more general theory, yet to be determined. Supersymmetry (SUSY) is one of the most promising SM extensions, addressing some of its shortcomings. SUSY predicts the existence of partners for SM particles, which cancel the SM corrections to the Higgs boson mass, solving the naturalness problem. In "natural" SUSY models, several particles are expected to be light and therefore observable at the LHC. First of all higgsinos, which share the same tree-level mass parameter as the Higgs boson. Then top squarks (stops), that provide a one-loop correction to its mass, and gluinos, that give a two-loop correction since they contribute at one-loop at the stop mass. The requirement on the stop mass reflects also on the sbottom mass, to which is related through the weak-isospin symmetry. These particles are exactly the target of the searches presented in this thesis. Furthermore, in the framework of R-parity conserving (RPC) SUSY, supersymmetric particles are produced in pairs and the lightest supersymmetric partner (LSP) is stable. In several models the LSP is neutral and weakly interacting, providing a good candidate for dark matter.

In this dissertation I discuss two searches for SUSY in final states enriched in *b*-jets, the collimated sprays of particles originating from the hadronization of a

© Springer Nature Switzerland AG 2020
C. Rizzi, *Searches for Supersymmetric Particles in Final States
with Multiple Top and Bottom Quarks with the Atlas Detector*, Springer Theses,
https://doi.org/10.1007/978-3-030-52877-5_1

b-quark. I have been strongly involved in both analyses, and in this thesis I describe with more emphasis the topics where I have given direct contribution.

The first analysis discussed is a search for gluino pair production, where each gluino decays through a stop or a sbottom to respectively four top or four bottom quarks and the LSP, which is assumed to be neutral and stable. Since the top quark decays to a *b*-quark and a *W*-boson, both the gluino decay chain through stop and sbottom lead to a final state with multiple *b*-jets and missing transverse momentum ($E_{\mathrm{T}}^{\mathrm{miss}}$).

The second search targets a general gauge mediation (GGM) model of higgsino pair production, where each higgsino then decays promptly to a Higgs boson and a gravitino, which in this case is the LSP. The high branching ratio of the Higgs boson into a pair of *b*-quarks makes a final state rich in *b*-jets promising to tackle this signature.

The content of this dissertation is organized as follows. Chapter 2 presents an introduction to the SM, moving then to its shortcomings and possible extensions, with particular focus on SUSY. Chapter 3 describes the LHC accelerator complex, the general techniques used in detectors for high-energy physics and the details of the ATLAS detector. Chapter 4 discusses the physics of *pp* interactions and how they are simulated with Monte Carlo (MC) techniques. The event reconstruction and the identification of the physics objects used in the analyses is presented in Chapter 5. Chapter 6 describes the main statistical procedures used to derive quantitative results in the analysis of the LHC data. Chapter 7 gives a general introduction to the strategies common to both the analyses discussed in this thesis, which are discussed in Chap. 8 and Chap. 9 for the gluino search and the higgsino search respectively. Chapter 10 presents a comparison of the analyses discussed in this thesis with other searches carried out by the ATLAS and CMS collaborations targeting similar signal models. Finally, the conclusions are discussed.

The results presented in this dissertation have lead to the following publications:

- ATLAS Collaboration, "Search for Supersymmetry in final states with missing transverse momentum and multiple *b*-jets in proton-proton collisions at $\sqrt{s} = 13$ TeV with the ATLAS detector", JHEP 06 (2018) 107.
- ATLAS Collaboration, "Search for supersymmetry in final states with missing transverse momentum and multiple *b*-jets in proton-proton collisions at $\sqrt{s} = 13$ TeV with the ATLAS detector", ATLAS-CONF-2018-041.
- ATLAS Collaboration, "Search for pair production of higgsinos in final states with at least three b-tagged jets in $\sqrt{s} = 13$ TeV pp collisions using the ATLAS detector", arXiv:1806.04030 [hep-ex], Submitted to: Phys. Rev. (2018).

While the paper mentioned above is the first ATLAS result for the signal model with higgsino pair production, the high cross-section for gluino pair-production made strong-production multi-*b* signals among the ones that were targeted since Run 1. I have been heavily involved in all the strong-production multi-*b* results since the beginning of Run 2, in particular:

- ATLAS Collaboration, "Search for pair production of gluinos decaying via stop and sbottom in events with *b*-jets and large missing transverse momentum in *pp*

collisions at $\sqrt{s} = 13$ TeV with the ATLAS detector", Phys. Rev. D 94 (2016) 032003.
- ATLAS Collaboration, "Search for pair production of gluinos decaying via top or bottom squarks in events with b-jets and large missing transverse momentum in pp collisions at $\sqrt{s} = 13$ TeV with the ATLAS detector", ATLAS-CONF-2016-052.

Beside physics analyses, I have also been involved in studies related to the performance of the ATLAS hadronic calorimeter: as discussed in Appendix D, I have evaluated the impact of a non-linearity in the response of the photonmultiplier tubes of the hadronic calorimeter on the measurement of the ATLAS luminosity. I have also carried out studies related to the identification of b-jets, performing studies on the b-tagging efficiency for jets with high transverse momentum and contributing to the development and validation of a tool to facilitate the use of truth-tagging; this technique allows to reduce the statistical uncertainty on samples of simulated events when requiring a high number of b-jets.

Chapter 2
Introduction to Standard Model and Supersymmetry

This chapter presents an introduction to the SM of particle physics, the theory that nowadays best describes the subatomic world. In Sect. 2.1 a general overview of the SM is given. Section 2.2 discusses the limitations of the SM, and some of the theoretical extensions proposed to overcome them. Finally Sect. 2.3 focuses on Supersymmetry, arguably one of the most promising of these extensions. Throughout this chapter (as well as in the rest of this thesis) we will use natural units; we will thus use energy units to describe masses, as the speed of light (c) and the Planck constant ($\hbar$) are set to unity.

2.1 The Standard Model of Particle Physics

The SM is a renormalizable gauge quantum field theory based on the group $SU(3) \times SU(2) \times U(1)$. It was developed in the second half of the 20th century [1–3], and since then the description that it gives of the elementary particles and of their interactions has been accurately tested by several experiments. Many experimental discoveries have been guided by the SM predictions, including the discovery of the top quark [4, 5] and up to the latest one, the observation of the Higgs boson at the LHC in July 2012 [6, 7].

2.1.1 Particle Content of the Standard Model

In the SM, particles are described as excitations of quantum fields. In the following paragraphs, we introduce the quantum field theory description of fermions and bosons.

© Springer Nature Switzerland AG 2020
C. Rizzi, *Searches for Supersymmetric Particles in Final States
with Multiple Top and Bottom Quarks with the Atlas Detector*, Springer Theses,
https://doi.org/10.1007/978-3-030-52877-5_2

Table 2.1 Fermion content of the Standard Model. Each particle is listed with its electric charge and mass [8]

Generation	Leptons			Quarks		
	Flavor	Charge	Mass (GeV)	Flavor	Charge	Mass (GeV)
1st	ν_e	0	$<2 \times 10^{-9}$	u	+2/3	2.2×10^{-3}
	e	−1	5.1×10^{-4}	d	−1/3	4.8×10^{-3}
2nd	ν_μ	0	$<2 \times 10^{-9}$	c	+2/3	1.27
	μ	−1	0.10566	s	−1/3	0.096
3rd	ν_τ	0	$<2 \times 10^{-9}$	t	+2/3	173.2
	τ	−1	1.77	b	−1/3	4.66

2.1.1.1 Fermions

Matter constituents are half-integer spin fields (fermions). Fermions are further divided into two categories based on the type of interaction they experience:

Leptons   which experience only the electromagnetic and weak interactions.
Quarks   which experience both the electroweak and the strong interaction.

Both leptons and quarks come in three generations, and conventionally the numbering of these generations follows an order of increasing mass. A summary of the SM fermions is presented in Table 2.1. While it is possible to observe free leptons, quarks exist only in bound states (hadrons); this is because of the confinement property of the strong interaction, discussed in Sect. 2.1.3. Hadrons built of three quarks have spin $\frac{1}{2}$ and are named baryons, while mesons are formed by two quarks and have integer spin.

The free Lagrangian of a fermion is given by:

$$\mathcal{L}_{free} = \bar{\psi} \left(i\gamma^\mu \partial_\mu - m \right) \psi \,, \tag{2.1}$$

where ψ is the fermion field, m its mass, γ are the Dirac matrices and ∂_μ is the four-momentum derivative.

2.1.1.2 Bosons

Particles with integer spin are referred to as bosons. In the SM, force carriers are described through spin-1 fields. The SM includes also a spin-0 particle, the Higgs boson. It is the interaction with the Higgs boson field that allows all the other elementary particles to acquire mass, as described in Sect. 2.1.4.

The Klein-Gordon Lagrangian governs the kinematics of spin-0 neutral particles:

$$\mathcal{L}_{\text{free}} = \frac{1}{2} \partial^\mu \phi \partial_\mu \phi - \frac{1}{2} m^2 \phi^2,$$

while in the case of charged particles (described through a complex field) the Lagrangian becomes:

$$\mathcal{L}_{\text{free}} = \partial^{\mu}\phi\partial_{\mu}\phi^* - m^2\phi\phi^*.$$

The two equations above describe scalar particles. In the case of a vector field A^{μ}, the expression of the Lagrangian is the following:

$$\mathcal{L}_{\text{free}} = -\frac{1}{4}F^{\mu\nu}F_{\mu\nu} + \frac{1}{2}m^2A^{\mu}A_{\mu}. \tag{2.2}$$

This is the Proca Lagrangian. In the case of a massless particle, this reduces to the Maxwell Lagrangian:

$$\mathcal{L}_{free} = -\frac{1}{4}F^{\mu\nu}F_{\mu\nu}, \tag{2.3}$$

where $F^{\mu\nu} = \partial^{\mu}A_{\nu} - \partial^{\nu}A_{\mu}$.

2.1.2 Interactions and Gauge Invariance

The SM describes all the interactions among elementary particles, except for gravity, for which nowadays no renormalizable quantum field theory is formulated. Table 2.2 presents a summary of the SM interactions and the properties of the corresponding force carriers. More details about the strong and electroweak interactions are given in the following sections.

The interaction terms in the SM Lagrangian are introduced by promoting an already existing global symmetry of the Lagrangian (θ) to a local one ($\theta(x)$) function of the space-time coordinates. In general, given a Lagrangian which is invariant under a symmetry group, the fields transform as:

$$\psi \rightarrow e^{ig\theta^k\tau^k}\psi,$$

where g is the coupling constant of the field ψ under the interaction and τ^k are the generators of the group and obey commutation relations:

Table 2.2 Interaction in the Standard Model. Here the different force carriers are listed, with their electric charges and masses [8]

Interaction	Carrier	Electric Charge	Mass (GeV)
Strong	Gluons (g)	0	0
Electromagnetic	Photon (γ)	$<10^{-27}$	0
Weak	W^+, W^-	$+1, -1$	80.385
	Z	0	91.1876

$$\left[\tau^i, \tau^j\right] = i f^{ijk} \tau^k .$$

In the equation above, f^{ijk} is the structure constant of the group and is always zero for Abelian groups. Promoting this global invariance to a local one ($\theta^k \to \theta^k(x)$) implies adding to the theory:

- A number of massless gauge fields W_μ^k equal to the number of generators of the symmetry group, that transform as

$$W_\mu^k \to W_\mu^k + \partial_\mu \theta^k + g f^{klm} W_\mu^l \theta^m .$$

- A covariant derivative:

$$D_\mu = \partial_\mu - i g \tau^k W_\mu^k ,$$

  that substitutes the standard derivative in the Lagrangian.
- A free Lagrangian for the vector fields as in Eq. 2.2, with:

$$F_{\mu\nu}^k = \partial_\mu W_\nu^k - \partial_\nu W_\mu^k + g f^{klm} W_\mu^l W_\nu^m .$$

Note that the last term, of second order in the field, is present only for non-Abelian symmetry groups, since it is proportional to the structure constant. This has non-trivial consequences as the second-order term leads to self-interaction among the gauge fields.

The SM is a theory invariant under $SU(3)_C \times SU(2)_L \times U(1)_Y$. Imposing local invariance under $SU(3)_C$ leads to the theory of strong interactions, while $SU(2)_L \times U(1)_Y$ is the symmetry whose breaking gives origin to the electroweak interactions.

2.1.3 Strong Interaction

Quantum chromodynamics (QCD) is the theory that describes strong interactions, based on the symmetry group $SU(3)_C$, where the subscript C refers to the color, the quantum number associated with these interactions. This can assume three possible values denoted with red, blue, and green. The observable states, hadrons, are color singlets, while quarks (anti-quarks) carry only one color (anti-color) charge. Since the symmetry group is non-Abelian, also the corresponding eight gauge bosons (gluons) carry a color charge (bi-color, with one color and one different anti-color) and therefore interact not only with quarks but also among themselves. Since $SU(3)_C$ is believed to be an exact symmetry, gluons are massless.

The renormalization of a gauge theory leads to the definition of running coupling constants, whose value depend on the energy scale at which they are evaluated. In QCD, at leading order the dependence of the coupling constant on the energy scale Q is given by:

$$\alpha_s(Q^2) = \frac{12\pi}{(11N_C - 2n_f) \log \frac{Q^2}{\Lambda_{QCD}^2}} \, ,\qquad (2.4)$$

where N_C is the number of colors, n_f is the number of quark flavors that are active (i.e. whose mass is lower than the energy scale) and Λ_{QCD} is the infrared cutoff scale that sets the limit of validity of the perturbative approximation. In QCD $N_C = 3$, so for $n_f < 16$ the coupling constant decreases with the increase of the energy scale of the process considered. This behavior has important consequences on the properties of the strong interaction:

- At high Q^2 (i.e. at small distances), α_s becomes small enough for the perturbative approximation to be correct. In this case quarks and gluons behave as free particles; this behavior is referred to as asymptotic freedom [9, 10].
- When the momentum transfer is small (i.e. at large distances) α_s is large; this gives rise to confinement: quarks cannot be observed as isolated particles, as it is not possible to extract individual quarks from hadrons. When the distance between two quarks is increased, the potential energy increases as well, up to the point when it is energetically more favorable to create from the vacuum a quark-antiquark pair and thus a new hadron is formed.
- In a collider experiment, quarks and gluons will create a collimated spray of hadrons, referred to as jet.

The evolution of the QCD coupling constant with the energy scale has been verified experimentally, as shown for example in Fig. 2.1.

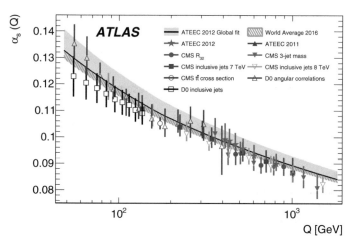

Fig. 2.1 Experimental determinations of the QCD coupling constant α_s. The 2016 world average is shown in the green hatched band. Figure from Ref. [11]

2.1.4 *Electroweak Interaction and Higgs-Englert-Brout Mechanism*

The theory of electroweak interactions is based on the symmetry group $SU(2)_L \times U(1)_Y$. This symmetry breaks at a scale around 100 GeV giving rise to the electromagnetic interaction, mediated by the photon, and to the weak interaction, mediated by the Z and $W^{\pm}$ bosons for neutral currents and charged currents respectively. The number of mediators, four, is the same as the number of generators of the symmetry group.

The $SU(2)_L$ part of the symmetry group governs the weak interactions, and the subscript L indicates that only left-handed particles participate to them. Left-handed and right-handed fields (ψ_L and ψ_R respectively) are defined through the chirality projectors P_L and P_R:

$$\psi_L = P_L\psi = \frac{(1 - \gamma_5)}{2}\psi \, ,$$

$$\psi_R = P_R\psi = \frac{(1 + \gamma_5)}{2}\psi \, ,$$

where γ_5 is defined as $\gamma_5 = i\gamma^0\gamma^1\gamma^2\gamma^3$. Looking back at Eq. 2.1, it can be noted that, if we decompose the fermion field into its left-handed and right-handed components, the derivative term keeps ψ_L and ψ_R separated, while the mass term mixes them:

$$-m\bar{\psi}\psi = -m\bar{\psi}P_L^2\psi - m\bar{\psi}P_R^2\psi = -m\bar{\psi}_R\psi_L - m\bar{\psi}_L\psi_R \, .$$

The covariant derivative for the $SU(2)_L \times U(1)_Y$ group is:

$$\mathcal{D}_\mu = \partial_\mu - ig'B_\mu Y - ig W_\mu^k T^k \, , \tag{2.5}$$

and substituting with this the regular derivative results in the interaction Lagrangian:

$$\mathcal{L}_{int}^{EW} = -\frac{g'}{2}\left(\bar{\psi}\gamma_\mu Y\psi\right)B^\mu - g\sum_k \left(\bar{\psi}\gamma_\mu T^k\psi\right)W_k^\mu \, ,$$

where we have introduced T^k, the weak isospin operator, and Y, the hypercharge operator (associated to the $U(1)_Y$ group), and the respective coupling constants g and g'. The quantum numbers of the T^k and Y operators relate to the electric charge Q through the Gell-Mann Nishijima relation:

$$Q = \frac{Y}{2} + T_3 \, ,$$

where Y is the hypercharge quantum number and T_3 is the quantum number of the third component of the isospin.

In the case of an $SU(2)$ symmetry, it is not possible to add directly to the Lagrangian a mass term for the vector bosons of the form in Eq. 2.2, as it would spoil the $SU(2)$ local invariance. The Higgs-Englert-Brout mechanism [12–14] solves this problem through spontaneous symmetry breaking (SSB) of the $SU(2)_L \times U(1)_Y$ invariance. The SSB is obtained by adding to the theory one extra isospin doublet of complex scalar components, the Higgs field:

$$\Phi = \begin{pmatrix} \phi^+ \\ \phi^0 \end{pmatrix} .$$

This doublet has hypercharge $Y = 1$ and isospin $T = \frac{1}{2}$; the first component has positive electric charge, while the second one is electrically neutral. The Lagrangian for this new field includes a kinetic and a potential term:

$$\mathcal{L}_\Phi = (\mathcal{D}_\mu \Phi)^\dagger (\mathcal{D}^\mu \Phi) - V(\Phi) , \qquad (2.6)$$

where $\mathcal{D}_\mu$ is the covariant derivative defined in Eq. 2.5 and the potential $V(\Phi)$ is given by:

$$V(\Phi) = \mu^2 \Phi^\dagger \Phi + \lambda (\Phi^\dagger \Phi)^2 .$$

The two real parameters μ^2 and λ relate respectively to the mass term and the strength of the self-interaction term. The shape of the potential depends on the value of these parameters:

- If $\lambda < 0$, the potential does not present any stable minima, and this case is therefore unphysical.
- If $\lambda > 0$ and $\mu^2 > 0$ there is only one solution to the minimization of the potential, $\Phi = 0$. This case is shown in Fig. 2.2a.
- If $\lambda > 0$ and $\mu^2 < 0$, the field acquires a vacuum expectation value (VEV) as the minima is not at zero; it lies instead on the points of the circumference such that:

$$\Phi^\dagger \Phi = \frac{\mu^2}{2\lambda} \equiv \frac{v^2}{2} . \qquad (2.7)$$

Figure 2.2b illustrates this case.

Up to this point the $SU(2)_L \times U(1)_Y$ symmetry is still intact, but the explicit choice of one of the infinite possible vacuum states for the Higgs field breaks the symmetry. According to the Goldstone theorem [15], the breaking of a continuous symmetry leads to the appearance of new massless scalar particles as field excitations. These new degrees of freedom are absorbed by the existing gauge bosons, thus giving them mass. To meet the experimental requirement of a massless photon, the choice of the Higgs vacuum has not to break the electromagnetic symmetry, $U(1)_{EM}$, so the component of the Higgs doublet that acquires a VEV is the neutral one:

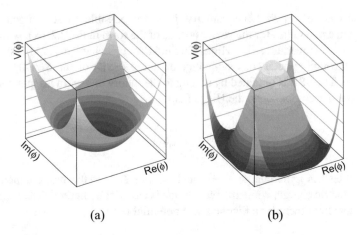

Fig. 2.2 Higgs potential in the case **a** $\lambda > 0$ and $\mu^2 > 0$ and **b** $\lambda > 0$ and $\mu^2 < 0$

$$\Phi_0 = \frac{1}{\sqrt{2}} \begin{pmatrix} 0 \\ v \end{pmatrix} \,.$$

We can then expand the field Φ considering small excitations around the minimum:

$$\Phi = \frac{e^{i\vec{\sigma}\cdot\vec{\theta}(x)/v}}{\sqrt{2}} \begin{pmatrix} 0 \\ v + \phi(x) \end{pmatrix} \,, \tag{2.8}$$

where $\phi(x)$ is the physical field associated with the Higgs boson, $\vec{\sigma}$ the Pauli matrices and $\vec{\theta}(x)$ the three degrees of freedom absorbed to give masses to the Z and $W^{\pm}$ bosons. Inserting Eq. 2.8 into Eq. 2.6 relates the mass of the Higgs boson to the parameters in the potential:

$$m_H = \sqrt{-2\mu^2} = \sqrt{2\lambda}v \,. \tag{2.9}$$

Since the numeric value of the μ^2 and λ parameters is not set, the theory does not predict a specific value for m_H. The physical mass eigenstates of the gauge bosons are a rotation of the interaction eigenstates, given by:

$$W_{\mu}^{\pm} = \frac{W_{\mu}^1 \mp W_{\mu}^2}{\sqrt{2}} \,,$$

$$A_{\mu} = B_{\mu} \cos\theta_W + W_{\mu}^3 \sin\theta_W \,,$$

$$Z_{\mu} = W_{\mu}^3 \cos\theta_W - B_{\mu} \sin\theta_W \,,$$

where we have introduced the Weinberg angle θ_W such that:

$$\tan \theta_W \equiv \frac{g'}{g} \ .$$

The $(\mathcal{D}_\mu \Phi)^\dagger (\mathcal{D}^\mu \Phi)$ term in Eq. 2.6 gives rise to the physical mass of the gauge bosons: once applied to the Higgs field, the covariant derivative in Eq. 2.5 produces terms quadratic in the gauge fields, that we interpret as mass terms:

$$\mathcal{L}_\Phi = \left(1 + \frac{\phi}{v}\right)^2 \left\{ m_W^2 \, W_\mu^\dagger W^\mu + \frac{1}{2} m_Z^2 \, Z_\mu Z^\mu \right\} + \mathcal{L}_H \ ,$$

where $\mathcal{L}_H$ denotes all the terms in $\mathcal{L}_\Phi$ that involve only the Higgs field: Higgs boson mass, cubic and quadratic self-interaction. Note that the coupling of the gauge bosons with the Higgs field is proportional to the square of the boson mass. At tree level the resulting masses of the gauge bosons are:

$$m_\gamma = 0 \ ,$$

$$m_Z = \frac{v\sqrt{g^2 + g'^2}}{2} \ ,$$

$$m_W = \frac{vg}{2} = \cos \theta_W m_Z \ .$$

Therefore, the Higgs-Englert-Brout mechanism generates automatically a mass term for the gauge bosons, that does not break the global underlying $SU(2)_L \times U(1)_Y$ symmetry. The Higgs field is also used to make fermion mass terms arise, but in this case it is necessary to postulate a Yukawa interaction between the Higgs and fermion fields. The fermion masses are assumed to be proportional to the strength of the coupling and, unlike the masses of the gauge bosons, they are not related to other parameters of the theory. While the Higgs field itself is enough to give mass to down-type fermions, the mass term for the up-type fermions requires the introduction of the complex conjugate of the Higgs field (Φ_C):

$$\Phi_C = i\sigma^2 \Phi^* = i \begin{pmatrix} 0 & -i \\ i & 0 \end{pmatrix} \begin{pmatrix} \phi^- \\ \phi^{0*} \end{pmatrix} = \begin{pmatrix} \phi^{0*} \\ -\phi^- \end{pmatrix} \ .$$

If we identify as y_f the coupling of the fermion f to the Higgs field, referred to as Yukawa coupling, the additional part of the Lagrangian that generates the fermion masses is:

$$\mathcal{L}_{Yukawa} = -\left[y_d \left(\bar{u}_L \ \bar{d}_L \right) \Phi d_R + y_u \left(\bar{u}_L \ \bar{d}_L \right) \Phi_C u_R \right] + h.c.$$

$$= -\frac{1}{\sqrt{2}} \left[y_d \left(v + \phi \right) \bar{d}_L d_R + h.c. + y_u \left(v + \phi \right) \bar{u}_L d_u + h.c. \right] \ .$$

In this equation we can now easily identify the fermion mass terms, of the form:

$$m_f = \frac{v}{\sqrt{2}} y_f \ .$$

Note that, in the case of fermions, the coupling to the Higgs boson is directly proportional to the fermion mass. The matrices y_f are not necessarily diagonal, but they can be diagonalized through a unitary transformation, which we can interpret as the transformation that relates the mass eigenstates to the weak interaction eigenstates. In the quark sector, this transformation is encoded in the Cabibbo-Kobayashi-Maskawa (CKM) matrix [16, 17], that describes the mixing of the down-type quarks. In the SM with three generations of fermions, the CKM matrix is parametrized by three angles and one complex phase that provides the only source of CP violation.

2.1.5 Measured Properties of the Higgs Boson

The value of the Higgs boson mass is not predicted by the SM, but it can be measured and, once it is know, the Higgs boson production cross-section and decay fractions can be predicted accurately. These predictions can be verified experimentally, supporting the idea that the discovered boson is indeed the SM Higgs boson. In 2015 the ATLAS and CMS collaborations published a combined measurement of the Higgs boson mass [18]:

$$m_H = 125.09 \pm 0.21 (\text{stat.}) \pm 0.11 (\text{syst.}) \ \ \text{GeV}.$$

This result uses the full Run 1 ATLAS and CMS data, analyzed in the $h \to \gamma\gamma$ and $h \to ZZ^* \to 4$ leptons channels.

The discussion of the Higgs mechanism in Sect. 2.1.4 highlights interesting properties of the interactions involving the Higgs boson. In particular, the strength of its coupling with fermions and bosons is proportional respectively to the mass and the square of the mass of the particle involved. This is reflected on the branching ratios (BRs): they are high for the decay to particles with the highest mass that are kinematically allowed. While the top quark mass is too large for the decay $h \to t\bar{t}$ to exist, the large top Yukawa coupling leads to loop-induced couplings of the Higgs boson to massless particles (gluons and photons); these are of phenomenological interest as they lead to the main production mode (gluon-gluon fusion) and to the $h \to \gamma\gamma$ decay mode, which has a low BR but leads to a very clean signature and was one of the key channels for the Higgs boson discovery. The theoretical production cross-section and BRs for a Higgs boson with mass between 120 and 130 GeV are reported in Fig. 2.3.

The production cross-section is found to be in agreement with the SM predictions within uncertainties. Figure 2.4a shows the best fit value of Higgs production cross-section times BR in the different production and decay modes [20]. Also the relation between the fermion or boson mass and the Higgs coupling has been verified experimentally, as show in Fig. 2.4b.

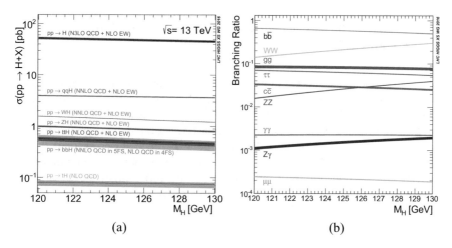

Fig. 2.3 **a** Higgs boson production cross-section at 13 TeV for a mass range between 120 and 130 GeV. **b** Higgs boson BR for a mass range between 120 and 130 GeV. Figures from Ref. [19]

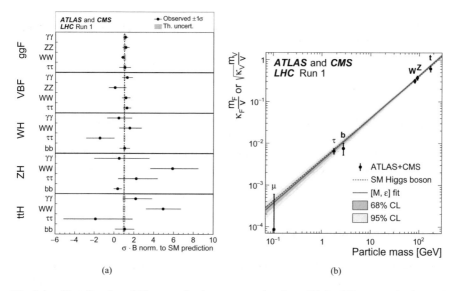

Fig. 2.4 **a** Best-fit value of Higgs production cross-section times BR in different production and decay modes. **b** Best-fit values as a function of particle mass for the combination of Run 1 ATLAS and CMS data. Figures from Ref. [20]

2.2 Limitations of the Standard Model and How to Extend It

Despite its undeniable success in describing the subatomic world, the SM has some limitations that suggest it should be considered as the low-energy approximation of a more general theory. Nowadays there is no perfect candidate to fill the role of this general theory, but the shortcomings of the SM highlight the characteristics this theory should have. Section 2.2.1 discusses the experimental observations that are not accounted by the SM framework, whose lacking is an objective limit of the SM. In Sect. 2.2.2 we review some other features that the SM is missing that, while not being strictly necessary, would be desirable in a general theory. Some theories and models candidate to extend the SM are briefly presented in Sect. 2.2.3.

2.2.1 Unexplained Phenomena

The SM lacks an explanation for some well established experimental phenomena that are listed in the next paragraphs.

2.2.1.1 Neutrino Masses

Neutrino oscillations [21] are possible only if there is a mass difference between the three neutrino generations, which automatically implies non-zero masses for at least some neutrinos. Although neutrino mass terms could be accommodated by the SM through right-handed neutrinos or a description of neutrinos as Majorana particles, the basic formulation of the SM describes neutrinos as massless particles.

2.2.1.2 Dark Matter and Dark Energy

The SM describes only baryonic matter; this accounts for about 5% of the energy density in the universe. Even though no direct observation of it has been made so far, we know there is also another type of matter, that does not have electromagnetic interactions and is about five times more abundant than ordinary matter. Since this type of matter does not reflect light, it is referred to as dark matter. The presence of dark matter has been postulated for the first time from the rotational velocity of galaxies [22], but this evidence has now been confirmed also by other observations, including the analysis of the cosmic microwave background from the WMAP and Planck collaborations [23, 24]. The sum of baryonic and dark matter accounts for about 32% of the of the energy density in the universe. The remaining 68%, is the energy responsible for the accelerated expansion of the universe, referred to as dark energy. While some extensions of the SM provide candidates for dark matter, at the moment no theory provides a compelling explanation for dark energy.

2.2.1.3 CP Violation

CP-violating processes are needed in order to generate the observed asymmetry between the matter and anti-matter content of our universe. While the SM provides one source of CP violation with the complex phase in the CKM matrix mentioned in Sect. 2.1.4, this is not enough and additional sources are needed in order to explain the observed asymmetry.

2.2.1.4 Gravity

Gravity is the only force acting on elementary particles that is not described by the SM. In fact, not only it is not described by the SM, but the simple attempt to quantize gravity through a spin-2 mediator (graviton) leads to a non-renormalizable theory. The strength of gravity is expected to be comparable to that of the other forces at the Planck scale ($\Lambda_{Planck}, \approx 10^{19}$ GeV).

2.2.2 Aesthetic Shortcomings

While the previous section discusses objective shortcomings of the SM, there are also some aesthetic criteria that the SM does not seem to satisfy. These are mostly based on the concept of naturalness: unless there is a good reason, the parameters of a "beautiful" theory should be all of the same order of magnitude, and a theory where instead the parameters are bound to assume very specific and different values (fine tuning) seems "unnatural". It is important to notice that this naturalness requirement is subjective, as well as the amount of fine tuning allowed for a theory to be considered natural.

2.2.2.1 Hierarchy Problem

The expression for the mass of the Higgs boson in Eq. 2.9 contains only the tree-level contribution. The proper computation should include the radiative contributions from all the particles that couple with the Higgs boson, directly or indirectly (see below). A fermion, whose Yukawa coupling leads to the interaction $-y_f \phi \bar{f} f$, will produce a correction to the Higgs boson mass with a divergent integral; this can be computed e.g. with a cut-off regularization, and in this case the resulting correction to the Higgs boson mass, shown in Fig. 2.5a, is:

Fig. 2.5 Correction to the Higgs boson propagator form the interaction with **a** a fermion and **b** a scalar

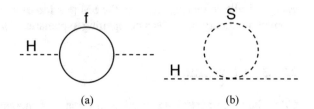

(a)   (b)

$$\Delta m_H^2 = -\frac{|y_f|^2}{8\pi^2}\Lambda_{UV}^2 + \mathcal{O}(\ln \Lambda_{UV}) , \tag{2.10}$$

where Λ_{UV}^2 is the cut-off scale, identified with the limit of validity of the theory. If we assume that no physics Beyond Standard Model (BSM) is present up to Λ_{Planck}, having corrections to the mass proportional to Λ_{Planck} requires a fine tuning of the parameters of the order of $\frac{m_H}{\Lambda_{Planck}} \approx 10^{-17}$. This is due to the strong hierarchy of the scales involved (hierarchy problem) [25–28]. Since the correction to the mass is proportional to the Yukawa coupling of the fermion, it is clear that the most important correction is the one given by the top quark, whose Yukawa coupling is $y_t \approx 1$.

Despite this, it can still be argued that the appearance of the Λ_{UV} divergence is connected more to the regularization scheme rather than to the theory itself. But even in this case, the value of 125 GeV for the Higgs boson mass remains difficult to justify. The SM Lagrangian does not have any symmetry that prevents the Higgs boson to couple to new BSM particles. If we assume the existence of a complex scalar that couples with the Higgs field through $-y_S|\phi|^2|S|^2$, the correction to the Higgs propagator, shown in Fig. 2.5b, is given by:

$$\Delta m_H^2 = \frac{y_S}{16\pi^2}\left[\Lambda_{UV}^2 - 2m_S^2 \ln(\Lambda_{UV}/m_S) + \cdots\right] . \tag{2.11}$$

Beside the first term, proportional to Λ_{UV}^2, that can be thought as a consequence of the cut-off regularization, we also have a second term proportional to m_S^2. If we assume that BSM physics exists and that, since it has not yet been observed, m_S must be large, this contribution drives m_H to high values. A similar argument applies even if the new BSM sector and the Higgs field do not couple directly but, for example, share a gauge interaction, which still gives rise to corrections proportional to the particle's mass through higher order loop diagrams.

2.2.2.2 Fermion Mass Hierarchy

Fermion masses are less sensitive than the Higgs boson mass to the cut-off scale as the divergence is only logarithmic. Nevertheless, it is striking how strong the fermion Yukawa coupling hierarchy is: even if we don't consider neutrinos, fermion masses span about six orders of magnitude, without any apparent reason.

2.2.2.3 Unification of Coupling Constants

The SM coupling constants of the $SU(3)_C$, $SU(2)_L$ and $U(1)_Y$ groups have a different evolution, and the extrapolation of their values at very high energies suggests a common value at a scale $M_{GUT} \approx 10^{15}$–10^{16} GeV, after which the different couplings should unify. However, if no new particles beyond the SM are involved in the computation of the evolution of the coupling constants, there is no exact crossing point, as shown in Fig. 2.6a.

2.2.2.4 Strong CP Problem

The most general SM QCD Lagrangian could include also a CP-violating angle, which would not spoil renormalizability. The presence of this term would have directly measurable physical consequences, for example a non-zero electric dipole moment for the neutron (nEDM). The tight limits on the nEDM ($<3.6 \times 10^{-26}$ e cm at 95% CL [29]) translate into upper limits on the parameter regulating the CP-violating term of the QCD Lagrangian ($\theta < 10^{-10}$ rad), while there is no theoretical reason why it should not be of order 1.

2.2.3 Extensions of the Standard Model

The aspects of the particle physics world not yet described by the SM are an indication that the SM should be extended. In this section we briefly discuss a few of the most popular models proposed to extend the SM, except from Supersymmetry which is discussed in more details in Sect. 2.3. These extensions introduce typically one or more of the following:

- New particles, which can either be independent particles or "partners" of the SM particles.
- New symmetries, which can be broken.
- New degrees of freedom, such as new quantum numbers or extra dimensions.
- New forces, whose strength depends on the energy scale.

2.2.3.1 Little Higgs

Little Higgs models address the problem of the Higgs boson mass by considering the Higgs boson as a pseudo-Goldstone boson of a global symmetry [30–33]. This leads to a light mass for the Higgs boson in a similar way as for the pion in QCD.

2.2.3.2 Technicolor

In technicolor models [25, 26] the mass of the Z and $W^\pm$ bosons is not generated through the interaction with the Higgs boson, but dynamically with a new asymptotically free gauge interaction whose strength is higher at small distances; the analogy with QCD (and the "color" degree of freedom) leads to the name technicolor. New fermions ("technifermions") are also introduced, transforming under the group vectorial representation. The Higgs boson is not predicted by technicolor, but it can be included in the model as a singlet scalar resonance, a dilaton, or a singlet pseudo-Goldstone boson.

2.2.3.3 Extra Dimensions

The first attempt to introduce additional spacial dimensions to unify forces was done in the 1920s by Kaluza and Klein [34, 35], who interpreted our four-dimensional universe as a "brane" of a higher-dimensional space-time. While the SM foresees only three spatial dimensions, adding extra dimensions can explain the observed weakness of gravity with respect to the other forces as it would be diluted in the extra dimensions. Particles propagating in the extra dimensions manifest in our brane as Kaluza-Klein modes. While the original model from Kaluza and Klein was disproved shortly after being proposed, similar ideas and formalisms have also been used afterwards:

- In the Arkani-Hamed-Dimopoulos-Dvali (ADD) model [36], two (or more) flat extra dimensions are added in which only gravity can propagate (through a graviton).
- In the Randall and Sundrum model [37] the universe is conceived as a five-dimensional Anti-de Sitter space-time, and the weakness of gravity is explained through the red-shift of the gravity brane.

2.3 Supersymmetry

Supersymmetry (SUSY) [38, 39] is an extension of the Poincaré group that rotates bosonic states into fermionic ones and vice versa, through the supercharge operator (Q) that carries itself a fermionic charge:

$$Q|\text{boson}\rangle = |\text{fermion}\rangle \,,$$
$$Q|\text{fermion}\rangle = |\text{boson}\rangle \,.$$

The Haag-Lopuszanski-Sohnius extension of the Coleman-Mandula theorem [40] allows such an operator as the only non-trivial extension of the Poincaré group in a consistent four-dimensional theory, if the operator and its hermitian conjugate ($Q^\dagger$)

satisfy the following relations:

$$\{Q, Q^\dagger\} = P^\mu \,,$$
$$\{Q, Q\} = \{Q^\dagger, Q^\dagger\} = 0 \,, \qquad (2.12)$$
$$[P^\mu, Q] = [P^\mu, Q^\dagger] = 0 \,,$$

where P is the four-momentum operator. These relations, which define the SUSY algebra, make it clear how the action of the supercharge is related to the Poincaré group: the combination of two SUSY rotations is a space-time translation.

2.3.1 Supermultiplets

The irreducible representations of the SUSY algebra are the supermultiplets, that contain both bosons and fermions. SUSY particles are referred to as superpartners of the SM fields within the same supermultiplet. Superpartners of the SM particles are indicated with the same symbol but with a $\sim$ on top of it. Depending on their particle content, supermultiplets are classified in different categories:

Chiral supermultiplets These supermultiplets are the ones containing the SM fermions and their superpartners. Each Dirac fermionic field can be regarded as two separate Weyl fields, each of which has two degrees of freedom and is associated with a complex scalar as a superpartner. The name given to superpartners of fermions is sfermions (e.g. the superpartner of the top is the stop and the superpartner of the bottom is the sbottom). Since chiral supermultiplets are formed by spin-0 and spin-$\frac{1}{2}$ particles, they contain also the SUSY extended Higgs sector, as described in Sect. 2.3.2.1.

Gauge supermultiplets Gauge bosons and their superpartners belong to gauge supermultiplets. Each spin-1 SM boson is associated to a spin-$\frac{1}{2}$ Weyl fermion, since before the electroweak symmetry breaking all the gauge bosons are massless and have only two degrees of freedom. The name of the superpartners of the gauge bosons is the same as the corresponding SM particle but with a -ino suffix (e.g. the gluon superpartner is the gluino).

Gravitational supermultiplets If we assume that gravity is mediated by the graviton, then a third type of supermultiplet is necessary that contains the spin-2 graviton and its superpartner, the spin-$\frac{3}{2}$ gravitino.

2.3.2 Minimal Supersymmetric Standard Model

SUSY is a framework that allows one to generate an infinite number of models. In the Minimal Supersymmetric Standard Model (MSSM) the number of extensions made to the SM is minimized, and is described by the superpotential:

Table 2.3 Chiral supermultiplets in the Minimal Supersymmetric Standard Model. The spin-0 fields are complex scalars, and the spin-1/2 fields are left-handed two-component Weyl fermions. Table from Ref. [41]

Names		Spin 0	Spin 1/2	$SU(3)_C,\ SU(2)_L,\ U(1)_Y$
Squarks, quarks ($\times 3$ families)	Q	$(\tilde{u}_L\ \ \tilde{d}_L)$	$(u_L\ \ d_L)$	$(\,\mathbf{3},\,\mathbf{2}\,,\,\frac{1}{6}\,)$
	$\bar{u}$	$\tilde{u}_R^*$	$u_R^\dagger$	$(\,\overline{\mathbf{3}},\,\mathbf{1},\,-\frac{2}{3}\,)$
	$\bar{d}$	$\tilde{d}_R^*$	$d_R^\dagger$	$(\,\overline{\mathbf{3}},\,\mathbf{1},\,\frac{1}{3}\,)$
Sleptons, leptons ($\times 3$ families)	L	$(\tilde{\nu}\ \ \tilde{e}_L)$	$(\nu\ \ e_L)$	$(\,\mathbf{1},\,\mathbf{2}\,,\,-\frac{1}{2}\,)$
	$\bar{e}$	$\tilde{e}_R^*$	$e_R^\dagger$	$(\,\mathbf{1},\,\mathbf{1},\,1\,)$
Higgs, higgsinos	H_u	$(H_u^+\ \ H_u^0)$	$(\tilde{H}_u^+\ \ \tilde{H}_u^0)$	$(\,\mathbf{1},\,\mathbf{2}\,,\,+\frac{1}{2}\,)$
	H_d	$(H_d^0\ \ H_d^-)$	$(\tilde{H}_d^0\ \ \tilde{H}_d^-)$	$(\,\mathbf{1},\,\mathbf{2}\,,\,-\frac{1}{2}\,)$

Table 2.4 Gauge supermultiplets in the Minimal Supersymmetric Standard Model. Table from Ref. [41]

Names	Spin 1/2	Spin 1	$SU(3)_C,\ SU(2)_L,\ U(1)_Y$
Gluino, gluon	$\tilde{g}$	g	$(\,\mathbf{8},\,\mathbf{1}\,,\,0\,)$
Winos, W bosons	$\tilde{W}^\pm\ \ \tilde{W}^0$	$W^\pm\ \ W^0$	$(\,\mathbf{1},\,\mathbf{3}\,,\,0\,)$
Bino, B boson	$\tilde{B}^0$	B^0	$(\,\mathbf{1},\,\mathbf{1}\,,\,0\,)$

$$
\begin{aligned}
W_{MSSM} &= y_u \tilde{u}\tilde{Q}H_u - y_d \tilde{d}\tilde{Q}H_d - y_e \tilde{e}\tilde{L}H_d + \mu H_u H_d \\
&= \sum_{F,G,T} y_u^{FG} \tilde{u}_F \tilde{Q}_G^T H_u^T - = \sum_{F,G,T} y_u^{FG} \tilde{u}_F \tilde{Q}_G^T H_u^T - \sum_{F,G,T} y_d^{FG} \tilde{d}_F \tilde{Q}_G^T H_d^T \\
&\quad - \sum_{F,G,T} y_e^{FG} \tilde{e}_F \tilde{L}_G^T H_d^T + \sum_T \mu H_u^T H_d^T \ .
\end{aligned}
$$

The superpotential appears in the MSSM Lagrangian through first and second order derivatives as discussed e.g. in Ref. [41].

The chiral supermultiplets of the MSSM are described in Table 2.3, and the gauge supermultiples in Table 2.4. Note that the states defined in these tables are interaction eigenstates, but not necessary mass eigenstates, as mixing is possible. In particular, the higgsino and the gaugino fields mix to form neutralino ($\tilde{\chi}_i^0$) and chargino ($\tilde{\chi}_i^\pm$) mass eigenstates. In addition, the sfermions associated with the left and right component of the same SM fermion mix to form the sfermion mass eigenstates (e.g. $\tilde{t}_L$ and $\tilde{t}_R$ mix to form $\tilde{t}_1$ and $\tilde{t}_2$). The gluinos instead do not undergo any mixing, as there are no other superpartners with the same quantum numbers. The gauge mass eigenstates of the MSSM are listed in Table 2.5, assuming that the mixing within the first- and second-generation sfermions is negligible.

Table 2.5 The gauge and mass eigenstates of the MSSM. Table from Ref. [41]

Names	Spin	P_R	Gauge Eigenstates	Mass Eigenstates
Higgs bosons	0	+1	$H_u^0 \ H_d^0 \ H_u^+ \ H_d^-$	$h^0 \ H^0 \ A^0 \ H^\pm$
Squarks	0	−1	$\tilde{u}_L \ \tilde{u}_R \ \tilde{d}_L \ \tilde{d}_R$	(same)
			$\tilde{s}_L \ \tilde{s}_R \ \tilde{c}_L \ \tilde{c}_R$	(same)
			$\tilde{t}_L \ \tilde{t}_R \ \tilde{b}_L \ \tilde{b}_R$	$\tilde{t}_1 \ \tilde{t}_2 \ \tilde{b}_1 \ \tilde{b}_2$
Sleptons	0	−1	$\tilde{e}_L \ \tilde{e}_R \ \tilde{\nu}_e$	(same)
			$\tilde{\mu}_L \ \tilde{\mu}_R \ \tilde{\nu}_\mu$	(same)
			$\tilde{\tau}_L \ \tilde{\tau}_R \ \tilde{\nu}_\tau$	$\tilde{\tau}_1 \ \tilde{\tau}_2 \ \tilde{\nu}_\tau$
Neutralinos	1/2	−1	$\tilde{B}^0 \ \tilde{W}^0 \ \tilde{H}_u^0 \ \tilde{H}_d^0$	$\tilde{\chi}_1^0 \ \tilde{\chi}_2^0 \ \tilde{\chi}_3^0 \ \tilde{\chi}_4^0$
Charginos	1/2	−1	$\tilde{W}^\pm \ \tilde{H}_u^+ \ \tilde{H}_d^-$	$\tilde{\chi}_1^\pm \ \tilde{\chi}_2^\pm$
Gluino	1/2	−1	$\tilde{g}$	(same)
Goldstino (gravitino)	1/2 (3/2)	−1	$\tilde{G}$	(same)

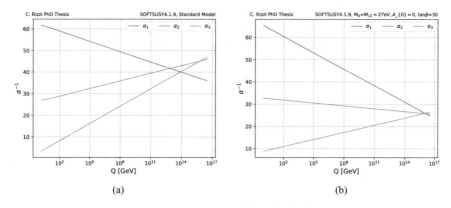

Fig. 2.6 Evolution of the gauge couplings with energy scale in **a** the SM and **b** the MSSM. Numbers obtained with SOFTSUSY [42]

Once the MSSM particles are introduced the computation of the evolution of the coupling constants, the unification of the forces discussed in Sect. 2.2.2 is achieved, as shown in Fig. 2.6b.

2.3.2.1 MSSM Higgs Sector

To maintain an holomorphic superpotential, SUSY requires at least two SU(2) doublets: one to give mass to up-type quarks (H_u) and one to give mass to down-type quarks (H_d). The two doublets have eight degrees of freedom. Three of them are needed to give masses to the gauge bosons, while the other five become observable particles:

- Two CP-even neutral Higgs bosons: h^0 and H^0, where h^0 is defined to be lighter than H^0.
- A^0, a CP-odd Higgs boson.
- Two charged Higgs bosons, H^+ and H^-.

2.3.2.2 R-Parity

The MSSM does not include all the possible terms that are gauge- and SUSY-invariant, to avoid interactions that would violate lepton or baryon number conservation. Matter parity, defined as:

$$P_M = (-1)^{3(B-L)} ,$$

is a multiplicative quantum number whose conservation implies automatically the conservation of lepton and baryon numbers (denoted with L and B respectively), without having to impose their conservation by hand. Particles in the same super-multiplet have the same matter parity: $P_M = -1$ for lepton and quark supermultiplets, while the Higgs and gauge supermultiplets have $P_M = +1$.

A more convenient way to express the same conservation rule is R-parity:

$$P_R = (-1)^{3(B-L)+2s} ,$$

where s is the spin of the particle. This is equivalent to matter parity as $(-1)^{2s}$ is conserved in all interactions where angular momentum conservation holds. The advantage of R-parity over matter parity is that for all SM particles $P_R = 1$ and for all SUSY particles $P_R = -1$. If R-parity is conserved, in a collider experiment SUSY particles are always produced in pairs, and each SUSY particle always has an odd number of SUSY particles in its decay products. As a consequence, the lightest supersymmetric partner (LSP) is stable; if the LSP is electrically neutral, it interacts with ordinary matter mainly through gravity and thus it provides a good candidate for dark matter.

While the consequences of R-parity conservation are appealing (especially the provision of a dark matter candidate), this is just an assumption and is not deduced from the theory.

2.3.3 Natural SUSY

Particles within the same supermultiplet share the same electric charge, isospin and QCD color, as Q and $Q^\dagger$ commute with the generator of the gauge transformations. The third equality in Eq. 2.12 implies that $[(P^\mu)^2, Q] = 0$. If we think of this as an operator applied to a supermultiplet, this implies that all the particles within the same supermultiplet should also have the same mass. The superpartners also bring a radiative correction to the Higgs boson mass, since they couple to the Higgs boson with the same coupling constant as their SM partners ($y_S = y_f = y$). If we consider the case of a fermion and a sfermion, according to Eqs. 2.10 and 2.11 the two contributions have opposite sign and the total correction is:

$$\Delta m_H^2 = \frac{y}{16\pi^2} \left[m_f^2 \ln(\Lambda_{UV}/m_f) - m_S^2 \ln(\Lambda_{UV}/m_S) \right] .$$

This correction cancels if each SM particle and its superpartner have the same mass. Unfortunately we know that, if SUSY exists, it must be a broken symmetry, since the superpartners do not have the same mass as the corresponding SM particles (otherwise they would have been observed already). Since the correction to the Higgs boson mass becomes larger with the increase of the mass difference between particles in the same supermultiplet, SUSY remains a solution to the hierarchy problem as long as this mass difference is reasonably small. The notion of "natural SUSY" refers to the class of SUSY models that allow to solve (or mitigate) the fine-tuning problem that affects the Higgs boson mass; this has been a very active topic of discussion both before the start of the LHC (see e.g. Refs. [43, 44]) and after (e.g. Refs. [45, 46])

Not all the superpartners have the same relevance in the contribution to the Higgs-mass corrections, and a SUSY model can be natural even if most particles in it are extremely heavy. The particles that should be light in a natural SUSY model are:

- higgsinos, whose tree-level mass is directly controlled by the μ parameter,
- stops, which contribute at one loop level to the Higgs boson mass, and
- gluinos, which contribute at two loops, since they give a one-loop correction to the stop mass.

2.3.4 SUSY Breaking in the MSSM

Superpartners of the SM particles have not been observed yet. This means that, if they exist, their mass must be larger than that of the corresponding SM particle, and thus SUSY must be a broken symmetry. The Lagrangian can therefore be written as:

$$\mathcal{L} = \mathcal{L}_{SUSY} + \mathcal{L}_{soft} ,$$

where $\mathcal{L}_{SUSY}$ is the SUSY-conserving part derived from the superpotential, while $\mathcal{L}_{soft}$ encloses the terms that break SUSY. The suffix "soft" means that we allow only soft SUSY-breaking terms in the Lagrangian, whose couplings have positive mass dimension, so that they vanish at very high mass scales. The inclusion of SUSY-breaking terms requires the addition of new particles and interactions to the MSSM. While there is no unambiguous way to modify the theory to do this (some examples of SUSY-breaking mechanisms are described in Sect. 2.3.5), it is possible to include a general parametric form of the SUSY-breaking terms in the Lagrangian:

- Gaugino masses, which in the MSSM correspond to the bino, wino and gluino mass terms;
- Non-holomorphic scalar squared masses, which add mass terms for squarks and sleptons and contribute to the Higgs potential;
- Holomorphic scalar squared masses, that in the MSSM can be present only in a term such as bH_uH_d;
- Scalar cubic couplings.

The necessary SUSY-breaking part of the Lagrangian adds many free parameters to the MSSM: in total the MSSM has 105 parameters in addition to the SM ones.

2.3.5   SUSY-Breaking Mechanisms

The SUSY-breaking terms of $\mathcal{L}_{\text{soft}}$ can be generated in models where SUSY is spontaneously broken. This is the case if, just like for the electroweak symmetry breaking, the Lagrangian is SUSY-invariant, but the vacuum state is not. As it happens with the Higgs mechanism, the spontaneous breaking of a symmetry implies the appearance of a massless Goldstone particle (goldstino, $\tilde{G}$) that, in the case of SUSY, has to be a neutral Weyl fermion in order to have the same quantum numbers as the supercharge Q. To be massless and annihilate the fermion mass matrix, the goldstino field has to be proportional to:

$$\tilde{G} = \begin{pmatrix} \langle D^a \rangle / \sqrt{2} \\ \langle F_i \rangle \end{pmatrix} , \qquad (2.13)$$

where D^a and F_i are respectively the auxiliary real bosonic and complex fermion fields of the SUSY Lagrangian, and $\langle \rangle$ denotes their vacuum expectation value. For Eq. 2.13 to be non-trivial, at least one of the D^a or F_i has to have a non-zero VEV. This is related to the two possible mechanisms that lead to SUSY breaking: via a D-term or an F-term. The Fayet-Iliopoulos mechanism [47] allows a non-zero VEV for an auxiliary field D by adding to the Lagrangian a term proportional to the auxiliary field itself, with a coupling constant with the dimension of a squared mass:

$$\mathcal{L}_{\text{FI}} = -\kappa D .$$

However with this mechanism it is difficult to properly attribute masses to all the MSSM particles.

Another possibility is to assign a VEV to an F-term, with the O'Raifeartaigh mechanism [48]. Since in the MSSM there is not a good candidate to be a gauge singlet with a non-vanishing VEV for an F-term, this implies the addition of at least one set of chiral supermultiplets, such that F_i is non-trivial for at least one of them. A realization of this is possible for example with three chiral supermulpliplets Φ_i and a superpotential of the form:

$$W = -k\Phi_1 + m\Phi_2\Phi_3 + \frac{y}{2}\Phi_1\Phi_3^2 .$$

The resulting scalar potential at tree level has a minimum for $\phi_2 = \phi_3 = 0$ and is independent of ϕ_1 (where ϕ_i are the scalar fields in Φ_i), while the one-loop corrections lead to a non-zero global minimum when also $\phi_1 = 0$. In the following we will consider only SUSY breaking originating from the O'Raifeartaigh mechanism.

When SUSY is treated as a local symmetry, it has to include gravity, and the resulting theory is called supergravity. The gravity mediator is the spin-2 graviton, and its superpartner the spin-$\frac{3}{2}$ gravitino. The gravitino can be interpreted as the gauge field of SUSY, and it incorporates the goldstino degrees of freedom upon spontaneous SUSY breaking. Therefore, the gravitino is also indicated with the same symbol as the goldstino, $\tilde{G}$. By dimensional arguments, we can see that the gravitino mass $(m_{3/2})$ has to be proportional to:

$$m_{3/2} \approx \frac{\langle F \rangle}{\Lambda_P} , \tag{2.14}$$

where $\langle F \rangle$ is the VEV of the SUSY-breaking F-term, as it has to vanish both if SUSY is an exact symmetry and if the Planck scale is extremely high (and thus gravity is negligible).

For any SUSY-breaking mechanism, the MSSM soft terms do not appear at tree level, but are generated through radiative corrections. SUSY breaking happens in a hidden sector, while the MSSM particles live in the visible sector, and experience the consequences of the spontaneous SUSY breaking happening in the hidden sector through interactions that couple both sectors. Depending on which interaction is driving the communication between the SUSY-breaking sector and the MSSM, we can distinguish three main mechanisms:

2.3.5.1 PMSB

If supergravity and the other new physics effects that arise at the Planck scale are responsible for the connection of the hidden sector with the MSSM sector, then the order of magnitude of the soft parameters in the MSSM is:

$$m_{\text{soft}} \approx \frac{\langle F \rangle}{\Lambda_P} . \tag{2.15}$$

This ratio is motivated by the observation that m_{soft} should vanish both in the limit where SUSY is an exact symmetry, as well as in the limit where the Planck mass is very large and the interactions mediated by gravity are not sizable anymore. We refer to this scenario as Planck-scale-mediated supersymmetry breaking (PMSB) [49–52]. In PMSB models, the gravitino mass has the same order of magnitude as the mass of the other superpartners (as we can notice by comparing Eqs. 2.14 and 2.15), and therefore it cannot be particularly light.

2.3.5.2 AMSB

If the supergravity couplings that mediate SUSY breaking in PMSB models are absent, SUSY can be broken through loop effects. In anomaly-mediated supersymmetry breaking (AMSB) models [53, 54] scalar and gaugino masses are generated through one-loop corrections arising from superconformal anomalies. These corrections are always present, but are dominant only in models that do not include tree-level terms. In the simplest AMSB model this leads to the prediction of negative squared masses for sfermions; one possible extension to the model that solves this problem is the addition of a scalar mass parameter m_0^2, common to all scalar squared masses.

2.3.5.3 GMSB

In the case of gauge-mediated supersymmetry breaking (GMSB) [55–60] the interactions mediating the connection between the hidden sector and the MSSM are the same gauge interactions of the MSSM itself. This requires the addition of a set of chiral supermultiplets (mediators) that interact both with the source of supersymmetry breaking and with the MSSM particles (through loop corrections to their masses, involving the MSSM gauge interactions). The mediators contribute at one loop level to the mass of gauginos, while they have a two-loop contribution to the mass of squarks and leptons. In both cases, the mass scale of the superpartners is:

$$m_{\text{soft}} \approx \frac{\langle F \rangle}{M_{\text{mediator}}} . \tag{2.16}$$

SM gauge bosons are not affected by the loop corrections induced by the mediators, since their mass is protected by the gauge symmetry. If we compare Eq. 2.16 with Eq. 2.14 we can see that, as long as the mass of the mediators is lower than the Planck scale, the gravitino can be significantly lighter than the rest of the superpartners. In most GMSB models the gravitino is the LSP. While we could expect the decay rate to a $\tilde{G}$ to be low because of the weakness of the gravitational interaction, this is not the case as the gravitino inherits the gauge interactions of the goldstino it absorbs during the spontaneous SUSY breaking. SUSY breaking introduces a large number of free parameters with respect to the SM. Most of these new parameters lead to flavor-changing interactions that are highly constrained experimentally. In models of gauge mediation, the interaction between the SUSY-breaking sector and the MSSM happens only through gauge interactions, which are flavor-blind, and therefore flavor-changing effects are suppressed automatically.

Table 2.6 List of free parameters in the pMSSM

Parameter	Description
$M_1, M_2\ M_3$	Gaugino mass parameters
$\tan \beta$	Ratio of the VEVs of the two Higgs doublets
M_A	Pseudoscalar Higgs boson mass parameter
μ	Higgsino mass parameter
A_t, A_b, A_τ	Third generation trilinear couplings
$m_{qL}, m_{uR}, m_{dR}, m_{lL}, m_{eR}$	First (and second) generation sfermion masses
$m_{q3L}, m_{tR}, m_{bR}, m_{\tilde{L}}, m_{\tilde{\tau}_R},$	Third generation sfermion masses

2.3.6 Phenomenological MSSM

The phenomenological Minimal Supersymmetric Standard Model (pMSSM) [61] reduces the number of free parameters in the MSSM through some assumptions motivated by experimental evidence. In particular:

- To avoid flavor changing neutral currents, the sfermion mass matrices and all the trilinear couplings are diagonal.
- The presence of CP-violating terms is restricted to only the terms present in the SM by imposing that all the new parameters are real numbers.
- The first two generations of sfermions are degenerate, i.e. the corresponding elements in the first and second generation have the same mass, to circumvent existing limits on the splitting between the first and second squark generations.
- Since the trilinear couplings give rise to amplitudes that are proportional to the corresponding Yukawa coupling, only the third generation ones are relevant and the others are set to zero.

These assumptions allow to reduce the number of free parameters from 105 to 19 (summarized in Table 2.6), making phenomenological analyses possible.

References

1. Glashow SL (1961) Partial symmetries of weak interactions. Nucl Phys 22:579
2. Weinberg S (1967) A model of leptons. Phys Rev Lett 19:1264
3. Salam A (1980) Gauge unification of fundamental forces. Rev Mod Phys 52:525. [Science 210:723 (1980)]
4. D0 Collaboration (1995) Search for high mass top quark production in $p\bar{p}$ collisions at $\sqrt{s} =$ 1.8 TeV. Phys Rev Lett 74:2422. arXiv:hep-ex/9411001 [hep-ex]
5. Collaboration CDF (1995) Observation of top quark production in $\bar{p}p$ collisions with the collider detector at fermilab. Phys Rev Lett 74:2626
6. ATLAS Collaboration (2012) Observation of a new particle in the search for the Standard Model Higgs boson with the ATLAS detector at the LHC. Phys Lett B 716:1. arXiv:1207.7214 [hep-ex]

7. CMS Collaboration (2012) Observation of a new boson at a mass of 125 GeV with the CMS experiment at the LHC. Phys Lett B 716:30 arXiv:1207.7235 [hep-ex]
8. Particle Data Group Collaboration, Patrignani C et al (2016) Review of particle physics. Chin Phys C 40:100001
9. Gross DJ, Wilczek F (1973) Ultraviolet behavior of non-Abelian gauge theories. Phys Rev Lett 30:1343
10. Politzer HD (1973) Reliable perturbative results for strong interactions? Phys Rev Lett 30:1346
11. ATLAS Collaboration (2017) Determination of the strong coupling constant α_s from transverse energy–energy correlations in multijet events at $\sqrt{s} = 8$ TeV using the ATLAS detector. Eur Phys J C 77:872. arXiv:1707.02562 [hep-ex]
12. Englert F, Brout R (1964) Broken symmetry and the mass of gauge vector mesons. Phys Rev Lett 13:321
13. Higgs PW (1964) Broken symmetries and the masses of gauge bosons. Phys Rev Lett 13:508
14. Higgs PW (1964) Broken symmetries, massless particles and gauge fields. Phys Lett 12:132
15. Goldstone J, Salam A, Weinberg S (1962) Broken symmetries. Phys Rev 127:965
16. Cabibbo N (1963) Unitary symmetry and leptonic decays. Phys Rev Lett 10:531. [648 (1963)]
17. Kobayashi M, Maskawa T (1973) CP violation in the renormalizable theory of weak interaction. Prog Theor Phys 49:652
18. ATLAS and CMS Collaborations (2015) Combined measurement of the Higgs Boson mass in pp collisions at $\sqrt{s} = 7$ and 8 TeV with the ATLAS and CMS experiments. Phys Rev Lett 114:191803 arXiv:1503.07589 [hep-ex]
19. LHC Higgs Cross Section Working Group Collaboration, de Florian D et al. Handbook of LHC Higgs cross sections: 4. Deciphering the nature of the Higgs sector. arXiv:1610.07922 [hep-ph]
20. ATLAS and CMS Collaborations (2016) Measurements of the Higgs boson production and decay rates and constraints on its couplings from a combined ATLAS and CMS analysis of the LHC pp collision data at $\sqrt{s} = 7$ and 8 TeV. JHEP 08:045 arXiv:1606.02266 [hep-ex]
21. Super-Kamiokande Collaboration (1998) Evidence for oscillation of atmospheric neutrinos. Phys Rev Lett 81:1562
22. Zwicky F (1937) On the masses of nebulae and of clusters of nebulae. Astrophys J 86:217
23. Larson D et al (2011) Seven-year Wilkinson Microwave Anisotropy Probe (WMAP) observations: power spectra and WMAP-derived parameters. Astrophys J Suppl 192:16 arXiv:1001.4635 [astro-ph.CO]
24. Planck Collaboration, Ade PAR et al (2014) Planck 2013 results. XVI. Cosmological parameters. Astron Astrophys 571:A16. arXiv:1303.5076 [astro-ph.CO]
25. Weinberg S (1976) Implications of dynamical symmetry breaking. Phys Rev D 13:974. [Addendum: Phys Rev D19:1277 (1979)]
26. Susskind L (1979) Dynamics of spontaneous symmetry breaking in the Weinberg-Salam theory. Phys Rev D 20:2619
27. Gildener E (1976) Gauge-symmetry hierarchies. Phys Rev D 14:1667
28. 't Hooft G (1980) Naturalness, chiral symmetry, and spontaneous chiral symmetry breaking. NATO Sci Ser B 59:135
29. Pendlebury JM et al (2015) Revised experimental upper limit on the electric dipole moment of the neutron. Phys Rev D 92:092003
30. Georgi H, Pais A (1975) Vacuum symmetry and the pseudo-Goldstone phenomenon. Phys Rev D 12:508
31. Kaplan DB, Georgi H (1984) SU(2) x U(1) breaking by vacuum misalignment. Phys Lett B 136:183
32. Kaplan DB, Georgi H, Dimopoulos S (1984) Composite Higgs scalars. Phys Lett B 136:187
33. Dugan MJ, Georgi H, Kaplan DB (1985) Anatomy of a composite Higgs model. Nucl Phys B 254:299
34. Kaluza T (1921) Zum Unitätsproblem der Physik. Sitzungsberichte der Königlich Preußischen Akademie der Wissenschaften (Berlin), Seite, pp 966–972
35. Klein O (1926) Quantum theory and five-dimensional theory of relativity. (In German and English). Z Phys 37:895. [Surveys High Energ Phys 5:241 (1986)]

36. Arkani-Hamed N, Dimopoulos S, Dvali GR (1998) The Hierarchy problem and new dimensions at a millimeter. Phys Lett B 429:263. arXiv:hep-ph/9803315 [hep-ph]
37. Randall L, Sundrum R (1999) Large mass hierarchy from a small extra dimension. Phys Rev Lett 83:3370
38. Wess J, Zumino B (1974) Supergauge transformations in four-dimensions. Nucl Phys B 70:39
39. Salam A, Strathdee JA (1974) Supersymmetry and nonabelian gauges. Phys Lett B 51:353
40. Haag R, Łopuszański JT, Sohnius M (1975) All possible generators of supersymmetries of the S-matrix. Nucl Phys B 88:257
41. Martin SP. A supersymmetry primer. arXiv:hep-ph/9709356 [hep-ph]. [Adv Ser Direct High Energy Phys 18:1 (1998)]
42. Allanach BC (2002) SOFTSUSY: a program for calculating supersymmetric spectra. Comput Phys Commun 143:305–331. arXiv:hep-ph/0104145 [hep-ph]
43. Barbieri R, Giudice G (1988) Upper bounds on supersymmetric particle masses. Nucl Phys B 306:63
44. Dimopoulos S, Giudice GF (1995) Naturalness constraints in supersymmetric theories with nonuniversal soft terms. Phys Lett B 357:573. arXiv:hep-ph/9507282 [hep-ph]
45. Papucci M, Ruderman JT, Weiler A (2012) Natural SUSY endures. JHEP 09:035 arXiv:1110.6926 [hep-ph]
46. Casas JA, Moreno JM, Robles S, Rolbiecki K, Zaldívar B (2015) What is a natural SUSY scenario? JHEP 06:070 arXiv:1407.6966 [hep-ph]
47. Fayet P, Iliopoulos J (1974) Spontaneously broken supergauge symmetries and goldstone spinors. Phys Lett B 51:461
48. O'Raifeartaigh L (1975) Spontaneous symmetry breaking for chiral scalar superfields. Nucl Phys B 96:331
49. Chamseddine AH, Arnowitt R, Nath P (1982) Locally supersymmetric grand unification. Phys Rev Lett 49:970
50. Barbieri R, Ferrara S, Savoy C (1982) Gauge models with spontaneously broken local supersymmetry. Phys Lett B 119:343
51. Ibáñez L (1982) Locally supersymmetric SU(5) grand unification. Phys Lett B 118:73
52. Hall L, Lykken J, Weinberg S (1983) Supergravity as the messenger of supersymmetry breaking. Phys Rev D 27:2359
53. Randall L, Sundrum R (1999) Out of this world supersymmetry breaking. Nucl Phys B 557:79. arXiv:hep-th/9810155 [hep-th]
54. Giudice GF, Luty MA, Murayama H, Rattazzi R (1998) Gaugino mass without singlets. JHEP 12:027. arXiv:hep-ph/9810442 [hep-ph]
55. Dine M, Fischler W (1982) A phenomenological model of particle physics based on supersymmetry. Phys Lett B 110:227
56. Alvarez-Gaume L, Claudson M, Wise MB (1982) Low-energy supersymmetry. Nucl Phys B 207:96
57. Nappi CR, Ovrut BA (1982) Supersymmetric extension of the SU(3) x SU(2) x U(1) model. Phys Lett B 113:175
58. Dine M, Nelson AE (1993) Dynamical supersymmetry breaking at low energies. Phys Rev D 48:1277
59. Dine M, Nelson AE, Shirman Y (1995) Low-energy dynamical supersymmetry breaking simplified. Phys Rev D 51:1362. arXiv:hep-ph/9408384 [hep-ph]
60. Dine M, Nelson AE, Nir Y, Shirman Y (1996) New tools for low-energy dynamical supersymmetry breaking. Phys Rev D 53:2658. arXiv:hep-ph/9507378 [hep-ph]
61. MSSM Working Group Collaboration, Djouadi A et al (1998) The minimal supersymmetric standard model: group summary report. In: GDR (Groupement De Recherche) - Supersymetrie Montpellier, France, 15–17 April 1998. arXiv:hep-ph/9901246 [hep-ph]

Chapter 3
LHC and ATLAS

The analyses presented in this thesis use the proton-proton (*pp*) collision data at a center-of-mass energy $\sqrt{s} = 13$ TeV collected by the ATLAS experiment in 2015, 2016 and 2017.

The ATLAS experiment is one of the four main experiments at the LHC at the European Organization for Nuclear Research (CERN). Section 3.1 of this chapter describes the LHC accelerator complex. This is followed by a general description of the detectors used in high-energy physics in Sect. 3.2. The ATLAS detector is discussed in Sect. 3.3.

3.1 The Large Hadron Collider

In this section we give a brief introduction to the LHC [1], at the moment the largest and most powerful particle accelerator in the world, hosted by CERN and in operation since September 2008. The first data for physics have been collected in the period between 2010 and 2013, referred to as Run 1, at the center-of-mass energy of $\sqrt{s} = 7$ and later $\sqrt{s} = 8$ TeV, delivering to ATLAS 5.5 fb^{-1} at $\sqrt{s} = 7$ TeV and 22.8 fb^{-1} at $\sqrt{s} = 8$ TeV. After a shutdown of two years, in 2015 the LHC started the Run 2 data taking at $\sqrt{s} = 13$ TeV, which will continue until the end of 2018. In 2015–2017 the LHC has delivered to ATLAS 93 fb^{-1} of *pp* collisions.

3.1.1 A Circular Hadron Collider

The LHC is a circular hadron accelerator, located in a 26.7 km long underground tunnel (with a depth ranging between 50 and 140 m) that was previously hosting the

© Springer Nature Switzerland AG 2020

C. Rizzi, *Searches for Supersymmetric Particles in Final States*
with Multiple Top and Bottom Quarks with the Atlas Detector, Springer Theses,
https://doi.org/10.1007/978-3-030-52877-5_3

Large Electron-Positron Collider (LEP), a CERN accelerator that was operational from 1989 to 2000. The LHC can accelerate protons up to a design center-of-mass energy of 13 TeV. Accelerating particles to very high energies is necessary both to study the structure of the particles themselves at smaller scales, and to create heavy states in collisions. Cosmic rays provide a source of particles with energies up to 10^7 times higher than what the LHC is capable of, but these extremely energetic rays are very rare, and is is not possible to modify the flux. Accelerators provide a well controlled flux of particles of a specific type in a specific location, and this allows the study of these particles with dedicated detectors.

A circular accelerator simplifies the acceleration of particles, as this can happen over several revolutions. When a charged particle travels on an orbit of radius r under the effect of a magnetic field B, its momentum p is given by:

$$p = 0.3rB, \tag{3.1}$$

where the momentum is expressed in GeV, B in Tesla and the radius of the orbit in meters. For a given magnetic field, a larger radius allows to reach higher energies.

The choice of a collider over a fixed-target experiment is motivated by the possibility of reaching a higher energy in center-of-mass of the system: while in a fixed-target experiment this is proportional to the square root of the energy of the incoming particle, in a collider it is the sum of the energies of the two beams.

Suitable particles for a collider experiment need to fulfill two criteria: they need to be charged, in order to be accelerated and guided through electric and magnetic fields, and they need to be stable enough not to decay before being used for collisions. These criteria effectively limit the choice to protons, electrons, their antiparticles and ions.

At the LHC it has been chosen to study collisions with protons and lead ions. Three types of collisions are studied: pp, lead-lead (Pb-Pb) and also proton-lead (Pb-p). The main reason to prefer protons over electrons is the energy loss that affects charged particles accelerated in a circular trajectory (synchrotron radiation), which decreases with the fourth power of the mass of the particle:

$$\frac{dE}{dt} \propto \frac{E^4}{m^4 r^2} .$$

The larger mass of a proton with respect to an electron leads to a decrease by a factor 10^{12} in the energy lost through syncrotron radiation. This choice comes with a price: proton-proton collisions lead to less clean events, with a lot of soft interactions covering the interesting hard interactions. Furthermore, the center-of-mass energy is unknown as the particles taking part in hard interactions are not the protons themselves but their constituents.

3.1.2 Magnet System

The LHC is not a perfect circumference: it is composed of eight arcs (sectors), where the magnetic system is located, and eight straight sections containing the resonant cavities, the four interaction points with and the detectors, the equipment for beam injection and extraction, and other instrumentation. Magnetic fields are used to govern the trajectory of particles. In the LHC there are more than nine thousand magnets, constructed from a superconducting alloy of niobium and titanium. About 150 tons of super-fluid helium at a temperature of 1.9 K are used to maintain the magnet system in the superconducting regime. Different types of magnets are necessary to achieve a proper control over the trajectory of particles.

3.1.2.1 Dipoles

Dipoles are used to create a vertical magnetic field, so as to bend the particles in the horizontal plane and thus give the dominant circular orbit. The LHC has 1232 dipoles, each 15 m long and providing a magnetic field of 8.3 T. The current necessary to achieve this strong magnetic field is 11.8 kA.

3.1.2.2 Quadrupoles

The LHC has 858 quadrupoles, used for beam focusing. A single quadrupole can focus the beam either in the vertical or the horizontal plane, but it causes a defocusing in the other plane; conventionally a quadrupole is denoted as focusing if it is oriented to focus in the horizontal plane. A combination of focusing and defocusing quadrupoles separated by some drift space (FODO lattice) is used to keep both planes focused, and gives rise to Betatron oscillations.

3.1.2.3 Higher-Order Magnets

Beside dipoles and quadrupoles, in the LHC there are about 600 higher-order magnets that are used to maintain a good beam quality; e.g. sextupoles are used to correct the spread in Betatron tune caused by the quadrupoles.

3.1.3 Resonant Cavities

While the orbit of particles is governed by the magnetic fields, longitudinal electric fields are used for acceleration. In the LHC the electric filed is provided by radiofrequency (RF) cavities. There are overall 16 RF cavities, eight per beam, hosted in four cryo-modules. Each cavity can provide an accelerating field of 5 MV/m, and oscillates with a frequency of 400 MHz. Since the electric field changes over time with the oscillations, particles passing through the same point of a RF cavity at different times experience a different voltage; this produces a non-trivial longitudinal dynamics, where particles oscillate around the ideal synchronous particle with changes in momentum and phase (synchrotron oscillations). If we define the slip factor η as the relative change in frequency in synchrotron oscillations with the relative change in momentum:

$$\eta = \frac{\Delta f/f}{\Delta p/p} \, ,$$

and the compaction factor α as the relative change in frequency in orbit length with the relative change in momentum:

$$\alpha = \frac{\Delta L/L}{\Delta p/p} \, ,$$

the following relation holds:

$$\eta = \frac{1}{\gamma^2} - \alpha \, ,$$

where γ is the Lorentz factor of the particle. This means that while the energy of the particle is low ($\eta > 0$) an increase in momentum leads to an increase in frequency, while it leads to a decrease in frequency for $\eta < 0$. At the transition energy, a previously stable synchrotron phase becomes unstable and vice versa; this requires a rapid change in RF phase. This situation is illustrated in Fig. 3.1a. For example, a particle corresponding to the phase point A1 will arrive in the RF cavity after one corresponding to the stability point P1, and will experiment higher voltage and increase in momentum; if $\eta > 0$ this increase in momentum will translate in an increase in frequency and the particle will, at the following revolution, arrive earlier, while if $\eta < 0$ the frequency will further decrease and the particle will be eventually lost. The transition energy in the LHC is 53 GeV, well below the injection energy of 450 GeV, so the LHC is always above transition.

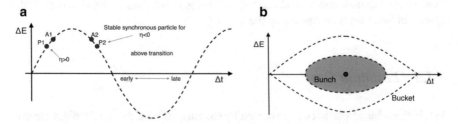

Fig. 3.1 a Phase stability below and above transition. **b** Bucket and the bunch for a beam above the transition energy. Figures based on the discussion in Ref. [2]

In the LHC beams particles are not distributed continuously, as this would not be allowed by phase instabilities, but are divided in bunches. The areas of stable motion are identified as bucket, and the area of the bucket is the beam longitudinal acceptance. The beam bunches fill only a part of the bucket, and the area of the beam bunches is the longitudinal beam emittance. Figure 3.1b shows a schematic view of the bucket and the bunch area for the case of a beam above the transition energy.

3.1.4 Luminosity and Operational Parameters

The amount of data delivered by an accelerator is quantified by the integrated luminosity $\mathcal{L}_{int}$. Given a certain process with production cross-section σ, the total number of events for that process is the product of cross-section and integrated luminosity:

$$N_{events} = \sigma\, \mathcal{L}_{int} \, .$$

The integrated luminosity is the time integral of the instantaneous luminosity $\mathcal{L}$,

$$\mathcal{L}_{int} = \int \mathcal{L}\, dt \, ,$$

which, assuming a Gaussian particle distribution and the same characteristics for the two beams, can be expressed as:

$$\mathcal{L} = \frac{f\, N_b n^2}{4\pi \sigma_x \sigma_y} \, , \tag{3.2}$$

where f is the revolution frequency, N_b the number of bunches in each beam, n the number of protons in each bunch and $\sigma_{x(y)}$ is the transverse beam size at the interaction point in the $x(y)$ direction.

The instantaneous luminosity defined in Eq. 3.2 needs to be corrected for two effects. First of all, this formula assumes a head-on collision between the bunches; in reality, to avoid unwanted interactions the beams collide with a crossing angle, and a large crossing angle decreases the instantaneous luminosity. The second correction is related to the beam transverse size, that can be expressed as:

$$\sigma_{x,y} = \sqrt{\epsilon \beta^*} \, ,$$

where ϵ is the beam emittance and β^* is the beta function. Beam collisions happen in minibeta insertions, drift spaces with a beta waist in the center, where the beam size is as small as possible. In the vicinity of the minimum, the beta function evolves like:

$$\beta(s) = \beta^* + \frac{s^2}{\beta^*} \, .$$

Table 3.1 LHC operational parameters in Run 2 compared to their design value. The numbers in parenthesis report changes in the parameters during the year. E.g. in 2016 the crossing angle was reduced after fill 5300, while for 2017 the numbers in parenthesis report the values after the recommissioning with $\beta^* = 0.3$ m

Parameter	2015	2016	2017	Design
Protons per bunch (n) [10^{11} p]	$\approx$1.2	$\approx$1.1	$\approx$1.2	1.15
Number of bunches (N_b)	2244	2220	$\approx$2250	2780
Emittance (ϵ) [mm mrad]	$\approx$3.5	$\approx$2.2	$\approx$2.2	3.5
Beta function (β^*) [cm]	80	40	40 (30)	55
Crossing angle [μrad]	290	370 (280)	300 (340)	285
Peak luminosity [10^{34} cm^{-2}s^{-1}]	0.51	1.4	1.7(1.9)	1.0

From this it is possible to see that the smaller the beta function, the larger the dependence with s, so a bunch with a finite size will not have the same beta function as a whole (hour glass effect). This correction becomes more important with the decrease of the β^* value. With the LHC design parameters, the effect of crossing angle and hour glass changes the instantaneous luminosity by about 20%.

The main limitations in the choice of the parameters that regulate the luminosity are collective effects, which could cause beam instabilities if the number of bunches or the number of protons per bunch is too high, and the limitations in the available aperture in the quadrupoles focusing the beam in the minibeta insertions, that impacts β^*. The summary of the LHC operational parameters during Run 2 is reported in Table 3.1.

The luminosity profile changes over time, and different experiments have different luminosity needs. ATLAS and CMS profit from having the maximum luminosity possible, while LHCb and ALICE have their top functionality at lower luminosity, and therefore apply a luminosity leveling that consists in changing the offset of the beams during the run, to maintain a constant (low) value of the instantaneous luminosity.

A high number of particles participating in a single bunch crossing can lead to a large number of multiple proton-proton collisions (pileup). In 2012 the LHC operated with a bunch spacing of 50 ns, leading to a pileup of about 35 events per bunch crossing. With the increase in energy from Run 1 to Run 2, the same settings would have implied more than 100 events, above the limit for the proper functioning of the detectors. This is what motivated the choice to move from the 50 ns bunch spacing used in Run 1 to the 25 ns in Run 2: with this setup, for the same instantaneous luminosity is possible to have a lower "per-bunch luminosity" and therefore a lower pile up. Figure 3.2 shows the luminosity-weighted distribution of the mean number of

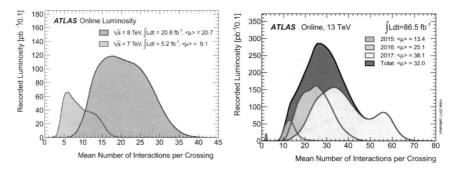

Fig. 3.2 Luminosity-weighted distribution of the mean number of interactions per crossing in **a** Run 1 and **b** Run 2 (2015–2017). All data recorded by ATLAS during stable beams is shown. The mean number of interactions per crossing μ corresponds to the mean of the Poisson distribution of the number of interactions per crossing calculated for each bunch. It is calculated from the instantaneous per-bunch luminosity as $\mu = \mathcal{L}_{bunch}\sigma_{inel}/f$, where $\mathcal{L}_{bunch}$ is the per-bunch instantaneous luminosity, σ_{inel} is the inelastic cross-section which is taken to be 80 mb for 13 TeV collisions, and f is the LHC revolution frequency. Figures from Ref. [3]

interactions per crossing during Run 1 (in Fig. 3.2a) and in 2015–2017 (in Fig. 3.2b) as recorded by the ATLAS experiment.

3.1.5 Accelerator Complex

Protons are injected in the LHC only after being accelerated to 450 GeV by a sequence of machines.

- Protons are extracted from H_2 at the Linac2 facility, a linear accelerator of 33 m that brings them to the energy of 50 MeV.
- The Proton Synchrotron Booster (PSB) is the first synchrotron in the acceleration chain (with a circumference of 157 m), that in 1.2 s increases the energy of the protons from 50 MeV to 1.4 GeV.
- With a circumference of 628 m, the Proton Synchrotron (PS) brings the protons to about 26 GeV. This was the oldest synchrotron experiment at CERN.
- The Super Proton Synchrotron (SPS) is the first accelerator of the chain to be underground (about 30 m) and has a circumference of 6.9 km; it brings the proton energy at 450 GeV. Beside preparing the protons to be injected into the LHC, the SPS provides beam also to the North Area, where the beams for fixed-target experiments are prepared, and to the AWAKE experiment, that studies proton-induced plasma wakefield acceleration.
- The LHC is the last step of this chain, and it accelerates the protons form 450 GeV to 6.5 TeV (the design energy is 7 TeV); in the LHC the protons gain about 0.5 MeV per turn, so it takes about 15 min to reach 6.5 TeV.

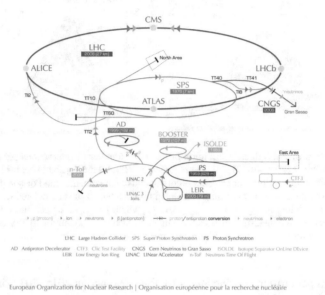

Fig. 3.3 Schematic view of the CERN accelerator complex. The four main LHC experiments are shown at the interaction points. Figure from Ref. [4]

The accelerator complex of CERN is shown in Fig. 3.3. The acceleration chain for lead ions differs from the one for protons in the initial part, which consists in Linac3 and Low Energy Ion Ring (LEIR) before the injection in the PS.

3.1.6   Experiments at the LHC

Seven experiments are built along the LHC circumference and collect the data produced during the collisions. Each experiment is run by an independent collaboration that comprises several universities and research institutes. ATLAS [5] and CMS [6] are the two largest and general purpose experiments, located at two opposite sides of the LHC ring, in correspondence with two of the four interaction points. Their goal is to study a large variety of SM processes and to perform an extensive search program for Beyond Standard Model (BSM) physics. The independent design of the two detectors, as well as of the separation between the two collaborations running them, is essential to provide a validation of the LHC results. Since the data used in this thesis is collected with the ATLAS detector, a more extensive description of this experiment is given in Sect. 3.3. The LHCb [7] and ALICE [8] experiments are located at the other two LHC interaction points. LHCb is a single-arm forward spectrometer, designed to perform high-precision studies of heavy flavor physics.

The ALICE experiment is dedicated to the study of $Pb - Pb$ collisions, which at the LHC happen with a center-of-mass energy of 2.6 TeV per nucleon pair; in this energy regime, quarks and gluons are expected to form a quark-gluon plasma. The position of the four main experiments on the LHC ring is shown in Fig. 3.3. Other three smaller experiments are installed along the LHC circumference: TOTEM [9], LHCf [10] and MoEDAL [11]. TOTEM is located at the same interaction point as CMS, and measures the total pp cross-section, as well as elastic and inelastic scattering. LHCf is installed in the same interaction point as ATLAS and has two detectors, at 140 m from each side of the collision point, aiming at the study of particles produced in the "forward" region (very close to the beam axis). MoEDAL is installed in the LHCb cavern and is designed to search for magnetic monopoles.

3.2 Detectors for Collider Physics

In this section we review the basic concepts that drive the design of the LHC detectors, including ATLAS.

3.2.1 Identification of Particles

The ability to accurately identify particles and reconstruct their energy and trajectory is what drives the design of detectors for high energy physics. In a detector, different sub-systems are able to capture different types of particle interactions, and the combination of the information collected by each of them allows to identify particles (or at least assign them to families, such as neutral or charged hadrons). A typical schema of the subdetectors sequence is shown in Fig. 3.4. The innermost layer, closer to the interaction point, is the tracking system, dedicated to the measurement of the signed charge and momentum of charged particles. The following layers are the electromagnetic and hadronic calorimeters, that measure the energy of particles with electromagnetic and hadronic interactions respectively. The outermost layer is dedicated to the muon system: because of their large mass (about 200 times more than electrons) muons do not produce electromagnetic showers and are therefore easy to identify as they are the only detectable particles that reach the external part of the detector.

3.2.2 Tracking and Spectrometry

A tracking device measures the traces left by charged particles passing through it. To allow the determination of the momentum and the charge of a particle, a tracking device needs to be accompanied by a magnetic field (in this case we speak of a

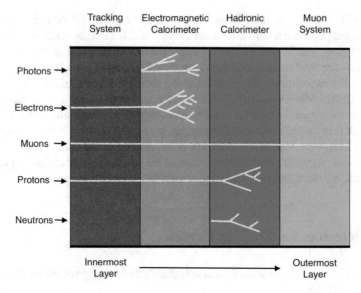

Fig. 3.4 Components of a typical detector for physics at colliders. Different particles are identified by the distinctive signatures in the subdetectors

magnetic spectrometer): once the magnetic field is known, the measure of the radius of curvature of the particle is equivalent to a measure of its momentum, according to Eq. 3.1. In a typical particle detector as the one in the schema in Fig. 3.4, both the inner tracking system and the muon system are magnetic spectrometers.

The relative uncertainty in the momentum is given by:

$$\frac{\sigma_p}{p} = \sqrt{(ap)^2 + b^2} \ . \tag{3.3}$$

The first term, whose relative importance increases for high-momentum particles, derives from the resolution in the measurement of the curvature. Typical values for a are between 0.01 and 1%. The constant term in Eq. 3.3 accounts for the impact of multiple Coulomb scattering, which broadens the distribution of the transverse momentum perpendicular to the direction of motion. This terms is important only at low energies, while it is negligible for high-energy particles.

There are three main configurations of magnetic fields typically used in momentum spectrometers:

- A dipole field leads to a rectangular symmetry; if we think of a circular collider with a coordinate system where z is the direction along the beam trajectory and (x, y) define a Cartesian system in the transverse plane, a dipole field in the x direction will cause a deflection in the (y, z) plane. This is the configuration adopted by forward spectrometers like LHCb, where the tracking devices are arranged in sequence in the z direction. As an example, the integral of the LHCb dipole field over the detector length is 4 Tm.

- A solenoidal field leads to a cylindrical symmetry and, if the field lines are along the z direction, the deflection is in the (x, y) plane. This is the typical configuration of the spectrometers in the central barrel, where the detectors are arranged in cylindrical layers. The CMS solenoid field is 4 T, while the ATLAS one is 2 T.
- A toroidal field leads to an azimuthal symmetry: the direction of the field lines is a circle in the transverse plane, and the deflection is in the (r, z) plane. This configuration is adopted by the ATLAS muon spectrometer, with a magnetic field of 4 T. As in the case of the solenoidal field, the detectors are arranged in cylindrical shells.

The track left by a charged particle curving because of the magnetic field is reconstructed by the tracking detectors. The two most common categories of tracking devices are gas and silicon detectors, described in the next two sections.

3.2.2.1 Gas Detectors

In gas detectors, the passage of a charged particle ionizes the atoms and creates electron-ion pairs. Once the electron and the ion are created, they can be separated (thus creating a current) by applying an electric field. This induces signal on an electrode added to the material, read through a readout system. The basic principle of most of gas detectors relies on a tube with a wire in the center, which is an anode with a high electric field. When the particle crosses the gas, the ionization electrons drift toward the anode and can be collected. Without an electric field, the electron and ion would move in the system by thermal diffusion. The effect of an electric field is to make the electron and ion move in opposite directions, allowing us to measure them. Since not many electrons are produced in gas, they need to be amplified. Inside the tube, the electric field decreases with the inverse of the distance from the anode wire. When the electrons reach a distance of a few micrometers from the wire, the electric field is very large and the electrons gain more energy than the ionization energy; this leads to secondary ionization and an exponential growth of the number of electron-ion pairs (avalanche effect).

Several types of gas detectors with different characteristics have been developed. In single-wire proportional chambers, like the one used in the outer tracker of the LHCb experiment, several counters are combined next to each other, to allow the measurement of the particle position. In multi-wire proportional chambers [12] several wires in two perpendicular directions are contained in the same box filled with gas. Drift chambers are a further evolution of wire chambers, where the position of the particle is computed by measuring the time taken by the secondary ionization to reach the anode. Time projection chambers (TPC) are a different type of gas detector, used for example in ALICE. It is constituted by a large area filled only with gas, without wires, with detectors and readout structure only at the end plates. Between the plates a strong electric field causes the electron-ion pairs to drift, until they reach the plates where a combined measurement of the position and time of arrival allows to reconstruct the three-dimensional trajectory of the particle (that can be curved if a magnetic field is added). Micro-strips chambers [13] are more modern, very condensed and thin gas detectors. Instead of being generated by a wire, the electric field comes from small metal deposits on a high-resistivity substrate.

In general the main problems of gas detectors based on ionization are the spatial extension and the long drift time, which nevertheless make them suitable for the outer part of the LHC detectors, in particular muon spectrometers.

All the gas detectors described above are based on the electron-ion pairs created by the ionization induced by the charged particle. A different principle is used in transition radiation detector (TRD). TRDs are based on detecting the electromagnetic radiation emitted by particles that cross boundaries between different media below the Cherenkov threshold [14]. The energy radiated increases with the energy of the incoming particle. These detectors are less common for tracking, but are mentioned here as the main example of a modern TRD in high-energy physics is the Transition Radiation Tracker (TRT) in the inner detector of the ATLAS experiment.

3.2.2.2   Solid-State Detectors

Semiconductor (and in particular silicon) detectors are the main type of solid-state detectors, and are currently the most used for inner tracking, where high precision and low occupancy are needed [15]. These detectors detectors are also based on the ionization of the material but in this case, since the structure is a crystal, we talk about electrons and electron holes.

The underlying principle is based on the band model of solids, describing the allowed energy levels for electrons in a solid: when many atoms of the same type are bound together in a crystal lattice, in order to fulfill Pauli's principle the atomic orbitals split into many closely spaced molecular orbitals, that can be considered as continuous energy bands. The highest-energy full band is the valence band, while the lowest partially-filled (or empty) band is the conduction band; the energy difference between the conduction band and the valence band is referred to as energy gap.

Semiconductors are materials where the conduction band is almost empty, but the energy gap is small (≈ 1 eV, e.g. 1.07 eV for silicon), so the conduction band can be occupied by excited electrons from the valence band; this leaves holes in the valence band, that under the effect of an electric field can drift as well. Semiconductors with a pure composition have the same amount of electrons and holes. If the semiconductor is doped with an atom with one more electron in the valence band, this creates an excess of electrons and the material is referred to as a n-type semiconductor. If instead the impurities are electron acceptors, the material has an excess of holes and is referred to as p-type semiconductor. When a p-type and a n-type semiconductors are pulled together, the holes of the p-side drift into the n-side, and the electrons from the n-side into the p-side. This creates an electric field that takes charge carriers out of the area where it is present (depletion region). When a charged particle passes through the semiconductor, it produces electron-hole pairs in the depletion region;

these drift apart because of the electric field and can be collected by electrodes. The application of a negative potential difference between the p-side and the n-side can increase the depletion area.

The density of silicon is about one thousand times higher than that of gases like argon, and silicon also needs a much lower energy to be ionized (3.6 eV for silicon, while it is about 250 times higher for argon). This leads to a much higher number of electron-holes pairs than the number of electron-ion pairs in gas detectors, so the signal needs very little amplification, and the size of the detector itself can be smaller.

Different configurations of the silicon sensors are possible. In a single-sided strip sensor, the readout strips are at negative potential. The second coordinate can be determined if also the n-side is divided into strips in a direction orthogonal to the ones in the p-side. The grid structure of this double-sided strip sensors can still lead to ambiguities when the number of hits is elevated; a true two-dimensional sensitivity is offered only by the pixel modules, where the module is divided in a matrix-like shape.

3.2.3   Calorimetry

Calorimeters can determine the energy of both charged and neutral particles through a destructive measurement: the energy of the particles is deposited in the detector material and transformed into a measurable quantity. Because of their sensitivity to a wide variety of particles, good energy resolution and relatively small size, they are very attractive devices for accelerator physics experiments [16, 17].

When the incident particle interacts with the material of the calorimeter it develops a cascade of particles (shower), with different characteristics for electromagnetic and hadronic interactions, described in the next two sections. Different types of calorimeters are necessary to capture the two typologies. The energy of the shower is decreased by the interactions happening in the absorber material, while the active material provides the conversion of the energy into a charge or light signal. In sampling calorimeters layers of absorber and active material are alternated in sequence, while in homogeneous calorimeters a single material carries out both functions.

3.2.3.1   Electromagnetic Calorimeters

The type of interaction that electromagnetically interacting particles have with the detector depends on their energy.

For photons and electrons with energies above 1 GeV, the dominant type of interaction is nuclear pair production and Bremsstrahlung, respectively. When these particles traverse a block of material they produce a cascade of particles (electromagnetic shower): electrons and positrons can emit a photon by Bremsstrahlung, and a photon, thanks to the interaction with a nucleus, can turn into an electron-positron pair.

The main parameter to describe the evolution of an electromagnetic shower is the radiation length (X_0), defined as the distance over which an electron reduces its energy to $1/e$ of the initial value, and it corresponds also to 7/9 of the mean free path for pair production for a photon. The radiation length depends on the characteristics of the material:

$$X_0 \left[\frac{g}{cm^2} \right] = \frac{716 \frac{g}{cm^2} A}{Z(Z+1) \ln \left(287/\sqrt{Z} \right)},$$

where A and Z are the atomic and mass number of the material. If we define $t = x/X_0$ as the shower depth relative to the radiation length, the maximum number of produced particles occurs at:

$$t_{max} = \frac{\ln (E_0/E_c)}{\ln (2)}.$$

Typical values for the radiation length are of the order of the cm (e.g. 0.56 cm for lead, 1.76 cm for iron [18]); 99% of the shower is contained in about 11(22) X_0 for a particle with an energy of 1 GeV(TeV), allowing for electromagnetic calorimeters of compact dimensions. The lateral width of the shower, determined mainly by multiple scattering, increases with depth and is defined in terms of the Molière radius:

$$R_M = \frac{21 MeV \ X_0 [\frac{g}{cm^2}]}{E_c[MeV]}.$$

A cylinder of radius $2R_M$ contains about 95% of the shower; for most calorimeters R_M has a value of few centimeters, so electromagnetic showers are quite narrow.

Once the electrons in the shower have an energy lower than the critical energy (E_c, defined as the energy where the loss through Bremsstrahlung equals the loss through ionization), the shower stops as the energy is dissipated mostly through ionization for electrons and photoelectric effect for photons, and no longer through the creation of new particles. Therefore all the energy of the incoming particle is in the end used to ionize the material of the detector, and this is the effect that is detected.

We have discussed in Sect. 3.2.2 how the resolution of the momentum measurement in a magnetic spectrometer decreases with the increase in the momentum itself. Instead, the relative energy resolution in a calorimeter improves for high-energy particles, and can be written in the parametric form:

$$\frac{\sigma_E}{E} = \sqrt{\left(\frac{a}{\sqrt{E}} \right)^2 + \left(\frac{b}{E} \right)^2 + c^2}.$$

The first term of the sum in quadrature reflects the stochastic nature of the shower development: ignoring the instrumental effects, the energy resolution of a calorimeter is proportional to the square root of the total track length, which is in turn proportional to the initial energy. The contribution of this term is small in homogeneous calorimeters, while is larger in sampling calorimeters (because of fluctuations in the fraction of energy deposited in the absorber) and it grows with the thickness of the absorber layers; typical values for a are 5–20% if the energy is expressed in GeV. The second term is the noise term originating from the electronic noise of the readout chain; this term is in general more relevant for calorimeters producing charge signals than for those producing light signals, and can become the dominant term for particles with energy below 1 GeV. The last term is a constant deriving from instrumental effects that produce a non-uniform detector response, including for example energy lost outside the detector volume and radiation damage; this becomes the dominant term at high energies and is typically <1%.

3.2.3.2 Hadronic Calorimeters

The difference between electromagnetic calorimeters and hadronic calorimeters finds its origin in the more complicated nature of strong interactions compared to the electromagnetic ones.

With respect to electromagnetic showers, hadronic ones have a much larger spatial extension. On the longitudinal direction, the scale is determined by the nuclear interaction length (λ_I), which is material-dependent and can be expressed as:

$$\lambda_I = 35 \frac{g}{cm^2} A^{1/3} ,$$

where A is the atomic number of the material. For most materials used in particle detectors this turns out to be larger than X_0 (e.g. the nuclear interaction length is 17.59 cm for lead, and 16.77 cm for iron [18]). The 99% shower containment is reached after about 5(9) λ_I for pions with E = 10(138) GeV. Also the lateral width of the shower is larger than in electromagnetic interactions: while the size of the Moliére radius is determined mainly by multiple scattering, the lateral profile of a hadronic shower depends on the transverse-momentum transfer, which can be quite sizable in strong nuclear interactions.

Another difference with electromagnetic showers lies in the composition of the shower: while an electromagnetic shower is constituted only by electrons, positrons and photons, a much larger variety of particles participates in hadronic showers, including both hadrons and electromagnetically-interacting particles. Starting from the simplifying assumption that one third of the particles produced in nuclear interactions are neutral pions ($f_{\pi^0} = 1/3$), a first approximation of the electromagnetic fraction of a shower is given by:

$$f_{em} = 1 - \left(1 - \frac{1}{3}\right)^n ,$$

where n is the number of generations in the shower. Since the number of generations increases with the initial energy, it is intuitive that also f_{em} will be larger for particles of higher energy. It is found [19]:

$$f_{em} = 1 - \left(\frac{E}{E_0}\right)^{k-1},$$

where E_0 is the energy necessary to produce one pion (e.g. 0.7 GeV for iron and 1.3 GeV for lead), and k is a slope parameter related to f_{π^0} through the average multiplicity $<m>$:

$$1 - f_{\pi^0} = <m>^{k-1}.$$

If we consider e.g. the particle spectra produced by protons with E = 100 GeV absorbed by lead, at low energies the particle content is dominated by electrons, positrons, photons and neutrons [16].

While in electromagnetic showers most of the initial energy is recorded in the detector, in hadronic showers a relevant fraction (up to 30–40%) is invisible. This invisible fraction is caused by energy that goes into breaking the nuclear bonds, nuclear fragments that in sampling calorimeters do not reach the active material, and neutral particles that can escape the calorimeter (e.g. neutrinos or long-lived neutral kaons). Therefore, for the same initial energy, the visible energy will be lower for a hadronic shower than for an electromagnetic one. If we define the response as the collected signal per unit of incident energy, the invisible energy causes a different response of calorimeters to the electromagnetic and to the purely-hadronic parts of the shower. Defining R_{em}, R_h, R_π respectively as the calorimeter response to electromagnetic shower, the hadronic part of the shower and to pions, we have that:

$$R_\pi = f_{em} R_{em} + (1 - f_{em}) R_h = R_h \left(\frac{R_{em}}{R_h} f_{em} + (1 - f_{em})\right).$$

In general, it is expected that $R_{em}/R_h > 1$. Since the value of the electromagnetic fraction is energy-dependent, the signal from the calorimeter does not increase linearly with energy. Compensating calorimeters are aiming at having the same response to the electromagnetic and hadronic part of the shower ($R_{em}/R_h = 1$), restoring the linearity of the calorimeter response. In non-compensating calorimeters the fluctuations in f_{em} are the dominant component of the energy resolution and, since the fluctuations are approximately Gaussian, they give rise to a term proportional to:

$$\frac{\sigma_E}{E} = \frac{\text{const}}{\sqrt{E}}.$$

In modern non-compensating calorimeters the value of the constant is about 0.4, while it can be as low as 0.2 in the case of compensating calorimeters.

3.2.4 Detecting Photons

As already mentioned in previous sections, photons are often the result of the passage of a particle through the material; this is the case e.g. in TRDs or in calorimeters with scintillator as active material. This sections gives an overview of how these photons can generate a detectable current; for a more extensive discussion see e.g. Refs. [20, 21]. The typical process of photon detection can be summarized in three different steps: (1) the incident photon generates a photoelectron or an electron-hole pair through the photoelectric or photoconductive effect, (2) the signal of the photo-electron is amplified, and (3) the amplified signal is collected. Important properties to classify photon-detecting devices are [18]:

Quantum efficiency: The probability that the incident photon generates a photo-electron. This depends on the photon wavelength.

Collection efficiency: The probability that the photoelectron is collected at the end of the chain.

Photon detection efficiency: The product of quantum and collection efficiency.

Gain: The amplification of the photoelectron, quantified as the number of electrons collected for each photoelectron generated.

Dark current or dark noise: The output current in the absence of signal.

Energy resolution: Resolution depending on electronic noise and statistical fluc-tuations.

Dynamic range: Maximum intensity that the detector can handle expressed in units of the smallest intensity with a signal-to-noise ratio above one.

Time dependence: Time between the arrival of the photon and the collection of the electrical current.

Rate capability: Inverse of the time needed after the arrival of a photon to be ready for another one.

Photon detectors can be broadly classified into three categories: vacuum, gaseous and solid-state detectors, described in the following paragraphs.

3.2.4.1 Photomultiplier Tubes

Photonmultiplier tubes (PMTs) are the most common type of vacuum photon detec-tors. A sketch of the structure of a PMT is shown in Fig. 3.5. It consists of a vacuum tube with an input window, a photocathode, a focusing electrode, electron multipli-ers, and an anode [22]. Photons enter the device through the window and excite the electrons in the valence band of the photocathode, which is a semiconductor; in transmission-type PMTs the photocathode is deposited on the inside of the window, while in reflection-type PMTs it is on a separate surface. The excited electrons dif-fuse to the surface of the semiconductor and, if they have enough energy to overcome the vacuum level barrier, are emitted as photoelectrons. These are accelerated and focused by the electrode toward the multi-stage dynodes, a system of electrodes coated with a secondary emissive material, where the incident electrons are mul-tiplied. The gain G of the PMT depends on the applied voltage V as $G = AV^{kn}$, where A and k are constants and n is the number of multiplicative stages. Typical values for the gain are 10^5–10^6. After the last stage of multiplication, the electrons are collected by the anode, which then outputs the current to the external circuit.

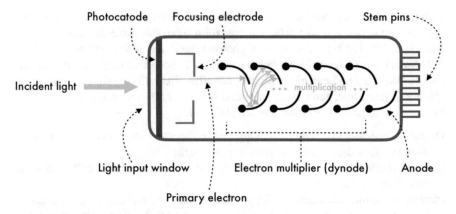

Photocatode   Focusing electrode   Stem pins

Incident light

Light input window   Electron multiplier (dynode)   Anode

Primary electron

Fig. 3.5 Sketch of the structure of a PMT

Beside PMTs, other examples of vacuum photon detectors are Microchannel plates (MCP) and hybrid photon detectors (HPD). MCPs are based on the same principle as PMTs but substitute the discrete multiplicative stages of the dynodes with continuous multiplication in cylindrical holes of a few μm. The decreased size of a MCP comes at the price of a large recovery time, shorter lifetime and smaller gain (typically $\approx 10^4$). In HPDs photoelectrons are accelerated onto a silicon sensor, allowing higher resolution in space and energy; this type of sensors are used e.g. in the CMS hadronic calorimeter.

3.2.4.2 Gaseous Photon Detectors

In gaseous photon detectors the photoelectrons are multiplied through the avalanche effect in an high-field region, similarly to what happens in gaseous tracking detectors. The photoelectrons are generated by the interaction of the photon either with a solid photocatode or with a photosensitive molecule vaporized and mixed in the gas itself.

3.2.4.3 Solid-State Photon Detectors

Solid state photon detectors are devices where the production and detection of the photoelectrons takes place in the same semiconductor material. They are in rapid development (see e.g. Ref. [23]) as they provide a smaller, and often cheaper, option to PMTs and gaseous detectors, especially when the area to be covered is small. Silicon photodiodes are devices based on a reverse-biased p-n junction, used in many applications from high-energy physics to solar cells. Photons passing through the silicon create electron-hole pairs through the photoconductive effect, which are then collected respectively at the positive and negative side of the chip.

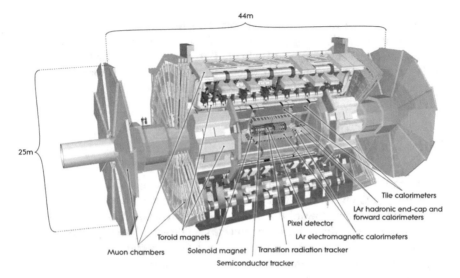

44m

25m

Tile calorimeters

LAr hadronic end-cap and
forward calorimeters

Pixel detector

LAr electromagnetic calorimeters

Toroid magnets

Muon chambers   Solenoid magnet   Transition radiation tracker

Semiconductor tracker

Fig. 3.6 Drawing of the ATLAS detector showing the different subdetectors and the magnet systems. Figure from Ref. [5]

3.3 The ATLAS Experiment

ATLAS [5], shown in Fig. 3.6, is the largest of the LHC detectors, measuring 44 m in length and 25 m in height, and weighting about 7000 tons. To be fully functional in the LHC environment, the ATLAS detector needs to be fast in order to resolve the collisions resulting from consecutive bunches (which are interspaced by 25 ns) and radiation resistant.

The ATLAS physics program covers a large variety of topics:

- SM processes can be measured at the LHC at energies never reached before, and being sensitive to them is essential both to provide accurate measurements and to use them as candles to calibrate the detector.
- The discovery of the Higgs boson was one of the main goals of the LHC and, after its observation in 2012, the focus moved onto measuring its properties.
- Compared to the Tevatron, the LHC is a true top-quark factory, and the study of the properties of this particle can both probe the SM and set limits on BSM theories.
- The LHC offers an exciting opportunity to discover BSM physics, and ATLAS needs to be ready to identify its signs.
- ATLAS has a dedicated program to study the properties of the physics involving b- and c-quarks, and of the physics of low mass states.
- ATLAS also carries out a heavy-ion program.

To cope with the wide range of types and energies (from few GeV to several TeV) of particles that need to be identified, ATLAS relies on a sequence of subdetectors nested in a cylindrical geometry, that follow the general schema discussed in Sect. 3.2.1: close to the interaction point (IP) we find the Inner Detector (ID), embedded in a solenoidal magnetic field of 2 T. The following layers are the electromagnetic calorimeter (ECal) and Hadronic Calorimeter (HCal), and the outermost part is occupied by the muon system, where muons are bent by a 4 T toroidal magnetic field. All these components are described in the next sections.

3.3.1 Coordinate System

ATLAS uses a right-handed coordinate system, with its origin at the nominal IP. The z-axis follows the beam direction, while in the transverse plane the y-axis points upward and the x-axis toward the LHC center. Positive and negative values of the z-axis identify respectively the A-side and the C-side of the detector. When spherical coordinates are used, the azimuthal angle ϕ is defined starting from the x-axis, and ranges between $-\pi$ and π; the polar angle θ is defined starting from the z-axis and takes values between 0 and π. The pseudorapidity is often used instead of the polar angle and is defined as:

$$\eta = -\ln \tan \left(\frac{\theta}{2} \right) .$$

In the limit of massless particles, is equivalent to the rapidity:

$$y = \frac{1}{2} \ln \left(\frac{E + p_z}{E - p_z} \right) ,$$

where E is the energy of the particle and p_z its momentum projected on the z-axis. The advantage of rapidity and pseudorapidity over the polar angle is that rapidity differences Δy are boost-invariant along the z-axis, as well as pseudorapidity differences $\Delta \eta$ for massless particles. The pseudorapidity is usually preferred to the rapidity as it does not require knowing the particle's mass but only its polar position. The η-ϕ plane is used to define the angular separation of two objects in the detector:

$$\Delta R = \sqrt{(\Delta \eta)^2 + (\Delta \phi)^2} . \tag{3.4}$$

Since protons are composite particles, and the hard scattering happens between its constituents, the longitudinal momentum of the partons is unknown. It is therefore useful to define the transverse momentum (p_T) as the projection of the momentum on the (x,y) plane:

$$p_T = \sqrt{p_x^2 + p_y^2} ,$$

where $p_{x(y)}$ are the projection of the momentum along the x-(y-)axis.

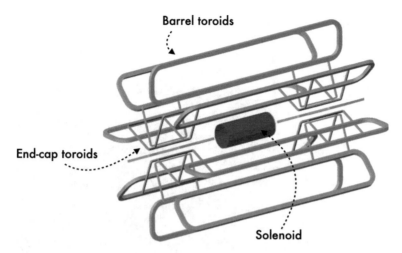

Fig. 3.7 Layout of the ATLAS magnet system. Figure adapted from Ref. [24]

3.3.2 Magnet System

The two segments of the ATLAS detector that are dedicated to tracking, the ID and the muon spectrometers, are embedded in two separate magnetic fields. A schematic overview of the ATLAS magnetic system is shown in Fig. 3.7.

As discussed in Sect. 3.2.2, the magnetic field configurations that are more suitable for a detector with cylindrical symmetry are solenoidal and toroidal. The bending of the charged particles in the ATLAS ID is caused by an axial 2 T solenoidal field, provided by the central solenoid [25]. This magnet is 5.8 m long, has an inner diameter of 2.46 m and an outer diameter of 2.56. The wounded coil is made of an Al-stabilized NbTi conductor, and is powered with a 7.73 kA current.

In the muon system, a solenoid would be disadvantageous in the measurement of forward muons, since the resultant magnetic field would not be perpendicular to the trajectory of those particles. The usage of a toroid to provide the outer magnetic field solves this problem. The choice of the "open air" toroid configuration allows a good muon reconstruction performance without relying on the ID. The toroids allow to efficiently generate the magnetic field over a large volume with a reduced amount of material. This minimizes the amount of multiple scattering, which represents one of the factors limiting the muon momentum resolution. The main drawback of the usage of a toroid is that, in order to obtain the same strength of magnetic field, a toroid needs more current than a solenoid (20.5 kA for 4 T). The ATLAS toroid system is divided into two subsystems, to allow for an easier design, as well as to provide access to the core part of the detector. The barrel toroid [26] provides a peak field of 3.9 T in the cylindrical shell between the calorimeters and the end of the muon spectrometer, and consists of eight coils contained in individual vacuum vessels. The end-cap toroids [27] provide the magnetic field necessary to bend the muons in the

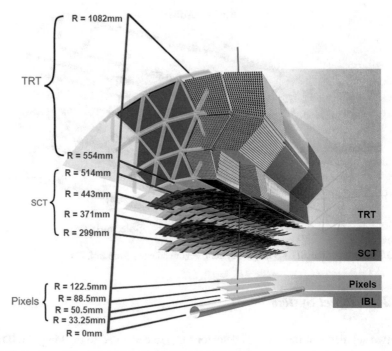

R = 1082mm

TRT

R = 554mm
R = 514mm
R = 443mm
SCT
R = 371mm
R = 299mm

TRT

SCT

R = 122.5mm
Pixels   R = 88.5mm
R = 50.5mm
R = 33.25mm
R = 0mm

Pixels
IBL

Fig. 3.8 Layout of the ATLAS ID. Figure from Ref. [30]

end-cap region of the spectrometer, and each of them consists of eight coils building
a single cold mass, originating a peak field of 4.1 T.

3.3.3   Inner Detector

The main purposes of the ATLAS ID [28, 29] are to provide a good momentum
resolution of the charged particles produced in the collisions and to allow the determination of secondary vertices. The dimensions of the ID are determined on one side
by the radius of the beam pipe and on the other side by the beginning of the ECal. The
total length is 5.4 m, which provides a coverage up to $|\eta| < 2.5$. The ID is divided
into a barrel, whose schema is shown in Fig. 3.8, and two end-cap regions, cover-
ing respectively the pseudorapidity regions $|\eta| < 1.2$ and $1.2 < |\eta| < 2.5$. In both,
a mixture of gaseous and silicon detectors is used to maximize the performance and
reduce the costs. The three components of the ID are the pixel detector, the semiconductor tracker, and the transition radiation tracker, discussed in the following
paragraphs.

3.3.3.1 Pixel Detector

The innermost layer of the ID is the pixel detector [31], divided into barrel and end-cap regions. During the Long Shutdown after the LHC Run 1, the ID was subject to important upgrades [30]. The main one is the addition of a fourth pixel layer in the barrel, in addition to the three already existing ones, the Insertable B-Layer (IBL) [32], that is positioned 3.33 cm away from the IP. In order to locate a detector so close to the IP, the beam pipe had to be replaced with a thinner one. The IBL is 72.4 cm long along the z-direction, and consists of 14 staves that provide full coverage in azimuthal angle. Each stave contains 20 modules, 12 with planar silicon sensors and eight with 3D pixel sensors [33], and each module has 144×328 pixels with an area of $50 \times 250 \, \mu m^2$ and a depth of 200 μm and 230 μm for the planar and 3D sensors respectively, for a total of over 12 million pixels in the entire IBL. The size of the pixels leads to a resolution of 40 and 8 μm respectively in the longitudinal and transverse direction. The three outer pixel layers are located at 5.05, 8.85 and 12.5 cm from the IP. Each module contains 80 pixels with an area of $50 \times 400 \, \mu m^2$ and a depth of 250 μm, leading to a spatial resolution of 115 and 10 μm respectively in the longitudinal and transverse direction. The two end caps, on the two sides of the detector, consist of three wheels each, with a radius of 34 cm and located at 49.5, 58.8 and 65.0 cm from the IP. To ensure a good performance, the pixel detectors need to be kept at a low and stable temperature, between -15 and $5 \, °C$ for the IBL, and between -15 and $-10 \, °C$ for the other layers.

3.3.3.2 Semi-conductor Tracker

The ATLAS Semiconductor Tracker (SCT) is composed by 4088 silicon micro-strip modules with binary readout mounted on carbon fibre composite structures, and is organized in four cylinders in the barrel and nine disks in each of the forward regions [34]. The cylinders have radii of 30.0, 37.3, 44.7 and 52.0 cm and provide a coverage for $|\eta| < 1.1 - 1.4$, while the disks cover the region with $1.1 - 1.4 < |\eta| < 2.5$. Out of the 4088 SCT modules, 2112 modules are in the barrel, and contain single-sided p-in-n silicon strips, with a pitch of 80 μm. In each module, the strip sensors are positioned back to back with an angle of 40 mrad, to be able to access information on the z-coordinate as well. The end-cap modules use strips with width between 56.9 and 94.2 μm. These choices lead to a spatial resolution of 580 μm in the longitudinal direction and 17 μm in the transverse one. Also the SCT components need to be kept at a low temperature, between -15 and $-5 \, °C$.

3.3.3.3 Transition Radiation Tracker

The TRT is the outermost layer of the ID. In the barrel it consists of 52544 straw tubes with a length of 1.5 m disposed parallel to the beam direction, while each end-cap contains 122880 straw tubes 0.4 m long disposed perpendicularly to the beam axis.

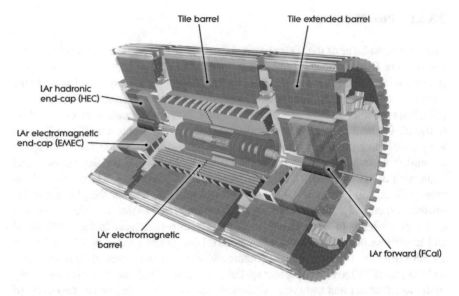

Fig. 3.9 Layout of the ATLAS calorimeter system. Figure from Ref. [5]

Each tube is 4 mm in diameter, and has in the inside gold plated tungsten wire as anode with a diameter of 31 μm. The tubes are filled with a mixture of 70% Xe, 27% CO_2 and 3% O_2; due to a gas leakage, in 2016 part of the TRT tubes have been filled with a cheaper mixture of 80% Ar and 20% CO_2. The TRT has a pseudorapidity coverage up to $|\eta| < 2$ and it provides tracking information only in the (r-ϕ) plane, with a resolution of 130 μm.

3.3.4 Calorimeters

The ATLAS calorimeter system is located outside the ID and the magnetic field of the solenoid, as shown in Fig. 3.9. The ECal is closer to the IP, while the HCal is on the outside; both systems have a barrel and an end-cap section. The combined thickness of the calorimeter system is about 11 interaction lengths to ensure the longitudinal containment of energetic jets, as well as a good reconstruction of the energy imbalance in the event, which is a measure of the energy carried away by neutral weakly-interacting particles. The total pseudorapidity coverage is up to $|\eta| < 4.9$.

Table 3.2 summarizes the energy resolution of the different subsystems. As expected from the discussion in Sect. 3.2.3, the resolution is better for electromagnetic showers than for the hadronic ones. For example, a 1 GeV particle detected in the ECal has an energy resolution of about 19%, while 50% (100%) if detected in the HCal barrel (end-caps). On the other hand, a 1 TeV particle has an energy resolution

Table 3.2 Energy resolution of the ATLAS calorimeters

Component	σ_E/E
ECal	$0.1/\sqrt{E[GeV]} \oplus 0.17/E \oplus 0.007$
HCal barrel	$0.5/\sqrt{E[GeV]} \oplus 0.03$
HCal end caps	$1/\sqrt{E[GeV]} \oplus 0.1$

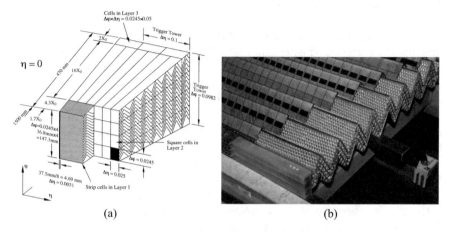

(a)   (b)

Fig. 3.10 **a** Schema of a barrel module of the ATLAS ECal. Figure from Ref. [5]. **b** Accordion shape of the metal plates of the ECal

of 0.7% in the ECal and 3% (10%) in HCal barrel (end-caps), and in this case the resolution is dominated by the constant term, related to instrumental effects and to the different response of the detectors to electromagnetic and hadronic showers.

3.3.4.1 Electromagnetic Calorimeter

The ECal is a sampling calorimeter with Liquid Argon (LAr) as active material and lead plates as absorber, both in the barrel and in the end caps. The lead plates have a characteristic accordion shape and, in the barrel, are oriented in the radial direction. Before the ECal, a presampler provides the information necessary to reconstruct the amount of energy lost in the passive material of the solenoid. The design of a LAr barrel module is shown in Fig. 3.10a, where it is possible to see the segmentation in three layers with decreasing granularity: the first layer is finely segmented in pseudorapidity, with strips of $\Delta\eta \times \Delta\phi = 0.0031 \times 0.098$; the second layer has towers of $\Delta\eta \times \Delta\phi = 0.025 \times 0.025$ to measure the clusters, while the third layer has wider towers of $\Delta\eta \times \Delta\phi = 0.05 \times 0.0245$ to provide an estimate of the energy leaking outside the ECal. The LAr barrel offers pseudorapidity coverage up to $|\eta| <$ 1.475, and the thickness of the detector varies from 22 X_0 at $\eta = 0$ to 33 X_0 at $|\eta| = 1.3$. Each of the two end-cap regions consists of two coaxial wheels, of eight modules each, that cover the region $1.375 < |\eta| < 3.2$, with a thickness varying between 26 and 36 X_0 for the inner wheel and between 24 and 38 X_0 for the outer wheel. The end-cap modules are divided into two layers, again with decreasing granularity.

3.3.4.2 Hadronic Calorimeter

The hadronic calorimeter is composed of three subsystems with different technologies: the Tile barrel calorimeter (TileCal) [35] is a sampling calorimeter with plastic scintillator as active material and steel as absorber, the hadronic end-cap calorimeter (HEC) uses copper as absorber and liquid argon as scintillator, while the forward calorimeter (FCal) also uses liquid argon in the active layer but has tungsten rods embedded in a copper matrix as absorber. The choice of the materials is driven by the need to have detectors more resistant to radiation in the forward region, where the flux of particles is larger.

TileCal covers the pseudorapidity region with $|\eta| < 1.7$, and is divided into a central long barrel (LB), 5.8 m long, and two extended barrels (EBs), 2.6 m long; the TileCal inner radius is 2.28 m, and the outer radius 4.25 m. Each barrel is divided in 64 modules, disposed on the ϕ direction and each having the size of 0.1 radians. Each module is further segmented radially into three layers with thicknesses of about 1.5, 4.1 and 1.8 λ_I in the LB and 1.5, 2.6 and 3.3 λ_I in the EB; a schematic of one module is shown in Fig. 3.11a. Ionizing particles passing through the plastic scintillator (polystyrene) produce ultra-violet light, which is then collected at the two edges of each tile and converted to the longer wavelength of visible light by wavelength-shifting fibers. The fibers, with a diameter of 1 mm each, transmit the light to the readout PMTs located in the grinder.

An approximately projective geometry, shown in Fig. 3.12, is provided by the grouping of the readout fibers into the PMTs: this defines a cell structure, and each cell has dimension $\Delta\eta \times \Delta\phi = 0.1 \times 0.1$ in the first two layers and $\Delta\eta \times \Delta\phi = 0.2 \times 0.1$ in the third layer. Special cells cover the gap region between the LB and the EB: the gap scintillators in the pseudorapidity region $1.0 < |\eta| < 1.2$ and the crack scintillators in the region $1.2 < |\eta| < 1.6$, in front of the LAr end caps.

The HEC shares the same cryogenic system as the ECal end caps, and covers the region with $1.5 < |\eta| < 3.2$. Liquid argon is more resistant to radiation than the plastic scintillator used in TileCal, and is therefore the preferred choice in the end-cap region. Each side of the HEC consists of two wheels with outer radius of 2.03 m, and each wheel is composed by 32 identical modules. The electromagnetic signal produced in the LAr is collected by cathodes on the plates.

The FCal provides coverage in the forward region with $3.1 < |\eta| < 4.9$. The FCal modules are located at high pseudorapidity, at a distance of 4.7 m along the z-axis from the IP.

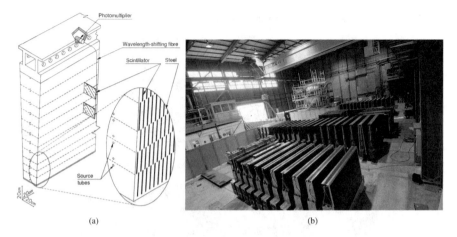

(a)   (b)

Fig. 3.11 **a** Schematic representation of a TileCal module and its interface with the optical readout. Figure from Ref. [5]. **b** TileCal modules before the installation

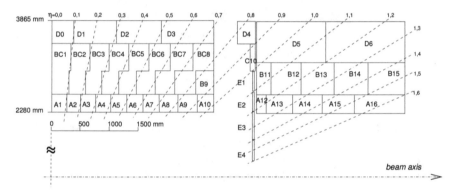

Fig. 3.12 Layout of the projective geometry of the TileCal cells. Figure from Ref. [5]

3.3.5 Muon Spectrometer

The muon spectrometer (MS) [36], shown in Fig. 3.13, is the outer layer of the ATLAS detector, and is located in the magnetic field produced by the 4 T toroidal magnets described in Sect. 3.3.2. It is designed to provide a p_T measurement with a relative uncertainty of 3% for muons of intermediate p_T, and to maintain a low uncertainty also at higher p_T (about 10% for muons with p_T of 1 TeV). It consists of four different muon chambers, two dedicated to the precise measurement of the muon tracks traversing the detector, and two providing fast event selection for the trigger system.

The two systems dedicated to precision measurement are the Monitored Drift Tubes (MDT) and the cathode strip chambers (CSC), both present in barrel and end caps. The MDTs are proportional drift chambers covering the region $|\eta| < 2.7$; they

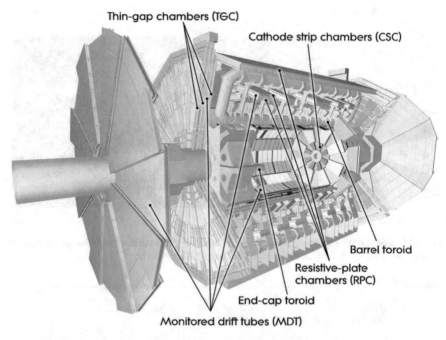

Fig. 3.13 Layout of the ATLAS muon system. Figure from Ref. [5]

are made of aluminium tubes with a diameter of 30 mm and a length between 700 and 6300 mm, and a cathode wire made of an alloy of tungsten (97%) and rhenium (3%). The filling gas is a mixture of Ar, N_2, and CH_4 with percentages respectively of 91%, 4% and 5%. The CSCs are multi-wire proportional chambers that cover the region with $2.0 < |\eta| < 2.7$, where the shorter drift time of this detector (30 ns compared to the 480 ns of the MDTs) allows to better cope with the increase in particle flux in the forward region. The anode wires in the CSCs are made of the same tungsten-rhenium alloy of the MDTs and have a 2.54 mm pitch, which is the same distance separating them from the copper cathode strips, creating a symmetric cell. The gas inside the chamber is a mixture of 30% Ar, 50% CO_2, and 20% CF_4. The spatial resolution is 40 μm in the bending plane and 5 mm in the perpendicular plane.

The muon trigger system needs to be able to identify events with energetic muons in a timescale compatible with assigning them to the correct bunch crossing, that are spaced by 25 ns. The two trigger chambers are the resistive plate chambers (RPC) and the thin gap chambers (TGC). The RPCs are arranged in three layers in the barrel region, outside the outermost MDT layer. Each narrow chamber consists of two parallel resistive bakelite plates and is filled with a mixture of 94.7% $C_2H_2F_4$, 5% Iso-C_4H_{10}, and 0.3% SF_6. The signal is read out by metal plates through capacitive coupling, providing a time resolution of 1.5 ns, while the space resolution is about 1 cm. The RPC also provides the ϕ coordinate for the track, which is not measured by the MDTs. The TGCs are multi-wire proportional chambers located in the end caps, filled with a mixture of 55% CO_2 and 45% $n-C_5H_{12}$. Contrarily to the CSCs, the TGCs are characterized by a distance between the cathode and the anode shorter than the anode pitch. The pseudorapidity coverage is $1.05 < |\eta| < 2.7$ for tracking and $1.05 < |\eta| < 2.4$ for triggering; the time response is similar to that of the RPCs, while the spatial resolution is better: between 2 and 7 mm.

3.3.6 Luminosity Measurement

An accurate determination of the luminosity is important both for SM measurements, where the luminosity uncertainty can dominate in some cases, and for BSM searches, where a precise background estimate is a key ingredient to be sensitive to a signal. In the ATLAS detector the luminosity measurement is performed with redundancy by multiple luminometers that use different technologies and algorithms, to allow a better determination of the final number and to assign systematic uncertainties. The instantaneous luminosity in Eq. 3.2 can also be expressed following the conventions in Ref. [37] as product of the number of bunch crossings N_b and the average luminosity per bunch cross $<\mathcal{L}_b>$:

$$\mathcal{L} = N_b <\mathcal{L}_b> = N_b \frac{f <\mu>}{\sigma_{\text{inel}}} . \tag{3.5}$$

With respect to the nomenclature of Eq. 3.2, $<\mu>$ is the average pileup per bunch crossing and σ_{inel} the inelastic pp cross-section. Because of the finite acceptance and efficiency of the detector, what is measured is:

$$\mathcal{L}_b = \frac{f <\mu_{\text{vis}}>}{\sigma_{\text{vis}}} ,$$

where $<\mu_{\text{vis}}>$ and σ_{vis} are the product of the corresponding quantities in Eq. 3.5 and the acceptance and efficiency of the detector. Out of these two quantities, $<\mu_{\text{vis}}>$ is directly measurable during the collisions, while σ_{vis} is determined with the van der Meer (vdM) method [38], carried out in the dedicated vdM runs. These are special runs with low bunch intensity and number of bunches, where a variation (scan) of the overlap of the two beams in the x- and y-direction is performed; the beam parameters and the peak of visible interaction rate per bunch crossing during the scan can be used to determine the visible cross-section and therefore calibrate each subsystem.

We can express the luminosity per bunch cross as:

$$\mathcal{L}_b = n_1 n_2 f \int \rho_1(x, y)\rho_2(x, y)\, dx\, dy , \tag{3.6}$$

where $\rho_1(x, y)$ and $\rho_2(x, y)$ are the particle densities in the two colliding bunches at the IP. Under the assumption that these densities can be factorized into the horizontal and vertical components, we can write Eq. 3.6 as:

$$\mathcal{L}_b = n_1 n_2 f \, \Omega_x(\rho_1(x), \rho_2(x)) \Omega_y(\rho_1(y), \rho_2(y)) \, ,$$

where $\Omega_{x/y}$ defines the beam overlap in the x/y direction.

The relevant quantity in physics analyses is the integrated luminosity over a defined period of time. The basic time unit over which the integrated luminosity is computed and stored is the luminosity block (LB), whose duration is defined by the ATLAS trigger system and is typically about one minute. The data contained in each LB is collected with the same detector conditions, and the integrated luminosity is computed as the average instantaneous luminosity multiplied by the LB time duration.

ATLAS has two primary specifically-designed luminometers, LUCID-2 (LUminosity measurements using Cherenkov Integrating Detector) and Beam Conditions Monitor (BCM), whose results are compared with the ones obtained by other ATLAS subsystems that measure luminosity through quantities that are sensitive to it, such as the number of tracks or the flow of particles.

3.3.6.1   Bunch-by-Bunch Luminometers

The two dedicated luminometers are able to provide information for individual bunch crossings, each labeled by a bunch-crossing identifier (BCID).

BCM consists of four 8×8 mm^2 diamond sensors, located at $z = \pm 184$ m from the IP and disposed in a cross shape around the beam pipe, at $|\eta| = 4.2$. Beside luminosity measurements, BCM also contributes to recognize beam losses so that the beam can be dumped before damaging the silicon detectors.

LUCID is located 17 m from the IP, at $5.6 < |\eta| < 6.0$, and consists on each side of 16 Cherenkov detectors built by aluminum tubes with a diameter of 10 mm. Cherenkov radiation is produced in the passage of particles through the quartz windows of the PMTs. A signal over threshold produces a hit for that bunch crossing.

3.3.6.2   Tracker-Based Algorithms

The ID, described in Sect. 3.3.3, records the passage of charged particles as tracks. The number of such charged tracks in each bunch crossing is proportional to the luminosity, and can be used as $<\mu_{vis}>$ if averaged over a LB. During collisions, partial data, which are selected with a random trigger that has the same probability of firing for each colliding bunch, are recorded in a dedicated stream. Because of the high number of colliding bunches, to achieve enough statistical precision the luminosity is provided as integrated over the LB; instead, during vdM scans, bunch-by-bunch luminosity can be provided. The reconstruction of the tracks used in this process is described in Sect. 5.1.

3.3.6.3   Bunch-Integrating Devices

The long-term stability of the luminosity measurement provided by BCM, LUCID and the track system is checked with devices that are sensitive to the flux of particles through the detector. This technique only allows to measure the instantaneous luminosity integrated over a time of a few seconds, and not bunch-by-bunch.

A subdetector capable of providing such measurement is TileCal, described in Sect. 3.3.4. The current generated by the PMTs is proportional to the total number of particles; this current is not read through the digital readout system, but through an integrator system sensitive to currents between 0.01 nA and 1.2 µA over a time window of 10–20 ms. The current induced in different cells varies largely with the cell position: the ones that receive a larger amount of particles are the ones around $|\eta| = 1.25$ and in the inner part of the detector (as the hadronic shower is stopped while it passes through the calorimeter). The TileCal luminosity measurement is not calibrated during vdM scans, but is instead equalized to LUCID or the track measurement for one specific run, and the calibration constant between current and luminosity allows to measure the luminosity also in different runs. More details on the TileCal luminosity measurement are given in Appendix D.

Additional systems used in the luminosity determination are the LAr-based calorimeters in the end caps, the electromagnetic one and the FCal. In both systems the high-voltage (HV) system maintains a constant voltage by supplying a continuous injection of current that counterbalances the voltage drop caused by the flux of particles in a certain sector. The measurement of this current provides a luminosity measurement. Also LAr systems are not calibrated during the vdM scan, but each HV run is calibrated to the baseline luminosity algorithm over a physics run.

3.3.7   Trigger System

With the nominal bunch spacing of 25 ns, the LHC produces collisions at a frequency of 40 MHz. This exceeds by several orders of magnitude the current capability to write events to disk; therefore the ATLAS trigger and data acquisition (TDAQ) system selects and records the events that are considered interesting for analyses. The TDAQ system has been updated between Run 1 and Run 2 [39] and it currently consists of a hardware Level 1 (L1) trigger and a single software-based high-level trigger (HLT). A schematic view of the Run 2 ATLAS TDAQ system is shown in Fig. 3.14. L1 is the first step in the chain, and reduces the output rate from 40 MHz to about 100 kHz. It uses reduced-granularity information from the calorimeter systems and from the muon RPC and TGS to select events with interesting objects and saves the information about the region of interest (RoI). L1 has a latency time of 2.5 µs; this corresponds to about 100 collisions, whose information has to be temporarily stored in buffers before the L1 Central Trigger Processor (CTP) finalizes a decision based on the inputs from the Level 1 Calorimeter trigger (L1Calo), Level 1 Muon trigger (L1Muon) and the Level 1 Topological trigger (L1Topo).

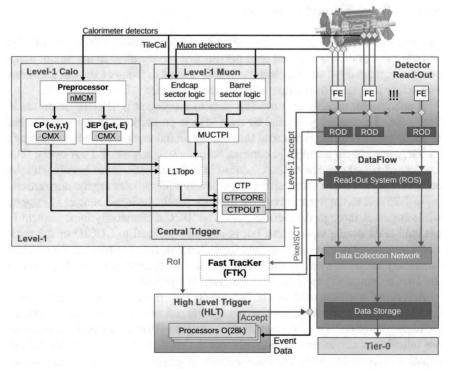

Fig. 3.14 Schematic view of the Run 2 ATLAS TDAQ system. Figure from Ref. [39]

During Run 1 the HLT consisted of two separate levels that in Run 2 have been merged to have a simpler setup and better resource sharing. The HLT uses information with finer granularity from the calorimeter systems, precision tracking from the muon spectrometer and tracking information from the ID, not available for L1. The HLT reduces further the event rate to about 1 kHz, and has a processing time of about 0.2 s per each event.

A trigger chain is a set of selections that characterize a certain trigger object, and the list of all available trigger chains defines the trigger menu; some of the items on the menu are unprescaled, which means all the events firing that trigger are stored, while others are associated to a prescale constant P such that only a fraction $1/P$ of the events firing the trigger are stored.

3.3.8 ATLAS Operation

ATLAS has been successfully recording the collision data delivered by the LHC in Run 1 and Run 2. Figure 3.15 shows the cumulative luminosity delivered by the LHC between the declaration of stable beams and the request to put the detector

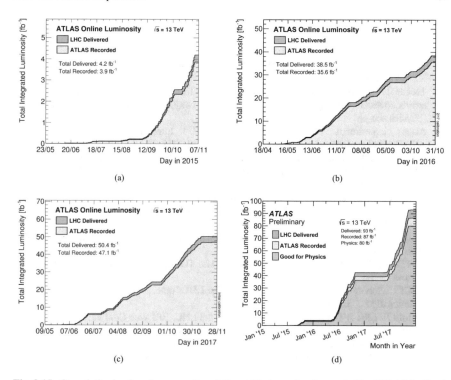

Fig. 3.15 Cumulative luminosity versus time delivered to (green) and recorded by ATLAS (yellow) during stable beams for pp collisions at $\sqrt{s} = 13$ TeV in **a** 2015, **b** 2016, and **c** 2017. **d** Cumulative luminosity versus time in 2015–2017 certified to be good quality data (blue). Figure from Ref. [3]

in standby for beam dump as well as the luminosity recorded by ATLAS in 2015, 2016 and 2017, which is the dataset used in the analyses described in this thesis. The difference between the delivered and recorded luminosity derives from inefficiencies of the data acquisition system and from the "warm start" procedure: the pixel and SCT high voltage and the preamplifiers of the pixel detector are turned on after the start of stable beams [3].

Not all the recorded luminosity can be used for physics analyses, as most of these require a good state of all the detector subsystems. This is checked for each luminosity block; the ones with poor detector conditions are disregarded, while the ones that pass this criteria form the good run list (GRL), which imposes requirements on the quality of the beams and of the detector. Figure 3.15d shows the total cumulative luminosity for the 2015–2017 period, highlighting the portion that enters the GRL.

References

1. Evans L, Bryant P (2008) LHC machine. J Instrum 3:S08001. http://stacks.iop.org/1748-0221/3/i=08/a=S08001
2. Tecker F (2014) Longitudinal beam dynamics. In: CAS - CERN accelerator school: advanced accelerator physics course: Trondheim, Norway, 18–29 Aug 2013. arXiv:1601.04901 [physics.acc-ph]
3. Atlas experiment, luminositypublicresultsrun2. https://twiki.cern.ch/twiki/bin/view/AtlasPublic/LuminosityPublicResultsRun2. Accessed 2018-02-18
4. Lefèvre C (2008) The CERN accelerator complex. Complexe des accélérateurs du CERN
5. ATLAS Collaboration (2008) The ATLAS experiment at the CERN Large Hadron Collider. JINST 3:S08003
6. CMS Collaboration (2008) The CMS experiment at the CERN LHC. JINST 3:S08004
7. LHCb Collaboration, Alves J, Augusto A et al (2008) The LHCb detector at the LHC. JINST 3:S08005
8. ALICE Collaboration, Aamodt K et al (2008) The ALICE experiment at the CERN LHC. JINST 3:S08002
9. TOTEM Collaboration, Anelli G et al (2008) The TOTEM experiment at the CERN Large Hadron Collider. JINST 3:S08007
10. LHCf Collaboration, Adriani O et al (2008) The LHCf detector at the CERN Large Hadron Collider. JINST 3:S08006
11. Pinfold JL (2014) The MoEDAL experiment at the LHC. EPJ Web Conf 71:00111
12. Charpak G, Bouclier R, Bressani T, Favier J, Zupancic C (1968) The use of multiwire proportional counters to select and localize charged particles. Nucl Instrum Meth 62:262
13. Oed A (1988) Position-sensitive detector with microstrip anode for electron multiplication with gases. Nucl Instrum Meth A 263:351
14. Ginzburg VL (1982) Transition radiation and transition scattering. Phys Scr 1982:182. http://stacks.iop.org/1402-4896/1982/i=T2A/a=024
15. Hartmann F (2009) Evolution of silicon sensor technology in particle physics. Springer Tracts Mod Phys 231:1
16. Fabjan CW, Gianotti F (2003) Calorimetry for particle physics. Rev Mod Phys 75:1243
17. Wigmans R (2000) Calorimetry: energy measurement in particle physics. Int Ser Monogr Phys 107:1
18. Particle Data Group Collaboration, Patrignani C et al (2016) Review of particle physics. Chin Phys C 40:100001
19. Gabriel T, Groom D, Job P, Mokhov N, Stevenson G (1994) Energy dependence of hadronic activity. Nucl Instrum Meth A 338:336
20. Rieke GH (2002) Detection of light, 2 edn. Cambridge University Press
21. Grupen C, Buvat I (eds) (2012) Handbook of particle detection and imaging, vols 1 and 2. Springer, Berlin, Germany
22. Hamamatsu (2017) Photomultiplier tubes: basics and applications. Hamamatsu photonics K. K., 3 edn. https://www.hamamatsu.com/resources/pdf/etd/PMT_handbook_v3aE.pdf
23. Renker D, Lorenz E (2009) Advances in solid state photon detectors. JINST 4:P04004
24. ATLAS Collaboration (2018) http://atlas-magnet-dfs.web.cern.ch. Accessed 2018-02-04
25. Yamamoto A et al (2008) The ATLAS central solenoid. Nucl Instrum Meth A 584:53
26. ATLAS Collaboration (1997) ATLAS barrel toroid: technical design report. https://cds.cern.ch/record/331065. CERN-LHCC-97-19
27. ATLAS Collaboration (1997) ATLAS endcap toroids: technical design report. https://cds.cern.ch/record/331066. CERN-LHCC-97-20
28. ATLAS Collaboration (1997) ATLAS inner detector: technical design report, vol 1. https://cds.cern.ch/record/331063. CERN-LHCC-97-16, ATLAS-TDR-4
29. ATLAS Collaboration (1997) ATLAS inner detector: technical design report, vol 2. https://cds.cern.ch/record/331064. CERN-LHCC-97-17

30. Potamianos K (2015) The upgraded Pixel detector and the commissioning of the Inner Detector tracking of the ATLAS experiment for Run-2 at the Large Hadron Collider. PoS EPS-HEP2015:261. arXiv:1608.07850 [physics.ins-det]
31. ATLAS Collaboration (2008) ATLAS pixel detector electronics and sensors. JINST 3:P07007
32. ATLAS Collaboration (2010) ATLAS insertable B-layer technical design report. https://cds.cern.ch/record/1291633. CERN-LHCC-2010-013, ATLAS-TDR-19
33. A. I. Collaboration (2012) Prototype ATLAS IBL modules using the FE-I4A front-end readout chip. J Instrum 7:P11010. http://stacks.iop.org/1748-0221/7/i=11/a=P11010
34. ATLAS SCT Collaboration, Jackson JN (2005) The ATLAS semiconductor tracker (SCT). Nucl Instrum Meth A541:89–95
35. ATLAS Collaboration (1996) ATLAS tile calorimeter: technical design report. Technical design report ATLAS. CERN, Geneva. https://cds.cern.ch/record/331062
36. ATLAS Collaboration (1997) ATLAS muon spectrometer: technical design report. https://cds.cern.ch/record/331068. CERN-LHCC-97-22, ATLAS-TDR-10
37. ATLAS Collaboration (2016) Luminosity determination in pp collisions at $\sqrt{s} = 8$ TeV using the ATLAS detector at the LHC. Eur Phys J C 76:653. arXiv:1608.03953 [hep-ex]
38. van der Meer S (1968) Calibration of the effective beam height in the ISR, Tech. Rep. CERN-ISR-PO-68-31. ISR-PO-68-31, CERN, Geneva. https://cds.cern.ch/record/296752
39. ATLAS Collaboration (2017) Performance of the ATLAS trigger system in 2015. Eur Phys J C 77:317 arXiv:1611.09661 [hep-ex]

Chapter 4
Proton-Proton Interactions and Their Simulation

Proton-proton interactions at the LHC are complex processes that span very different energy scales. In order to interpret the experimental data it is essential to develop a good understanding of the physics involved in proton-proton (*pp*) collisions. The ability to simulate the various processes is crucial to compare the observed data with the theory predictions. Section 4.1 focuses on the description of our understanding of a *pp* collision, while Sect. 4.2 discusses the event-simulation process; the main MC generators used in the ATLAS Collaboration are described in Sect. 4.3, and Sect. 4.4 touches briefly on the topics of detector simulation and data-driven corrections.

4.1 Proton-Proton Interactions

In hard-scattering processes, where the momentum transfer is much higher than the proton mass [1], a *pp* collision is easier to understand in terms of interactions between the constituents of the protons, quarks and gluons, collectively referred to as partons. A schematic view of a *pp* event is shown in Fig. 4.1. In this particular example, the interaction between a quark and a gluon leads to a final state with a *Z* boson and jets.

As can be appreciated from the figure, the hard process, which can be computed in perturbation theory, takes place between two of the proton's partons; the probability that a gluon or a specific quark type takes part in the hard scattering is related to the parton distribution functions (PDFs), discussed in Sect. 4.1.1. If the products of the hard scattering are quarks or gluons, they first of all loose energy by radiating other gluons (which in turn can generate quark-antiquark pairs through gluon splitting) in the process of parton shower; successively they evolve to stable hadrons in the lower-energy hadronization process, which we can describe only through phenomenological models. This picture is further complicated by the fact that also initial-state quarks and gluons can radiate. Also, the other partons not contributing

© Springer Nature Switzerland AG 2020
C. Rizzi, *Searches for Supersymmetric Particles in Final States
with Multiple Top and Bottom Quarks with the Atlas Detector*, Springer Theses,
https://doi.org/10.1007/978-3-030-52877-5_4

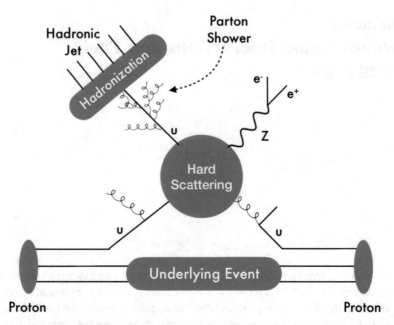

Fig. 4.1 Schematic representation of a *pp* collision, involving a quark-gluon scattering that leads to a final state consisting of a Z boson and a hard jet

to the hard scattering can interact, originating what is referred to as the underlying event.

4.1.1   Factorization Theorem

The hard scattering between the partons inside the proton takes place in a kinematic regime where the strong coupling constant, α_s, is small and therefore the partonic cross-sections can be computed in perturbation theory. Thanks to the factorization theorem [2], the generic production cross-section for a final state X can be expressed in terms of the partonic cross-section $\hat{\sigma}$ as:

$$\sigma(pp \rightarrow X) = \sum_{i,j} \int dx_1 dx_2 \, f_i(x_1, \mu_F^2) \, f_j(x_2, \mu_F^2) \, \hat{\sigma}_{ij \rightarrow X}(x_1 x_2 s, \mu_R^2, \mu_F^2).$$

(4.1)

The i and j indexes run over all possible partons, and $f_i(x_1, \mu_F^2)$ is the PDF for the parton of type i, representing the distribution of probability for that parton to carry a fraction x_1 of the proton momentum when the proton is probed at a scale μ_F (factorization scale). The partonic cross-section, $\hat{\sigma}_{ij \rightarrow X}$, is computed at the partonic center of mass energy $\sqrt{\hat{s}}$; it has to be noted that $\sqrt{\hat{s}}$ is lower than the total center of mass energy, as $\hat{\sigma} = x_1 x_2 s$, where x_1 and x_2 are the fraction of the proton momentum that is carried by each of the two partons. Although the partonic cross-section depends on μ_F and on the renormalization scale μ_R, which is the scale used for the evaluation of α_s, when considered at all orders in perturbative QCD this dependence disappears. Higher-order calculations exhibit a reduced scale dependence, and are therefore used whenever available.

4.1.2 Parton Density Functions

The partons inside the proton cannot be observed as free particles, and therefore their PDFs cannot be computed with perturbative QCD. In particular, for a given scale, it is not possible to predict theoretically the probability distribution of the parton's momentum fraction. Instead, once the PDFs are known at a certain scale, their energy evolution is determined by the equations derived independently by Dokshitzer [3], Gribov and Lipatov [4], and Altarelli and Parisi [5] (DGLAP equations):

$$\frac{\partial q(x, Q^2)}{\partial \log Q^2} = \frac{\alpha_s}{2\pi} \left(P_{qq} \otimes q + P_{qg} \otimes g \right),$$

$$\frac{\partial g(x, Q^2)}{\partial \log Q^2} = \frac{\alpha_s}{2\pi} \left(\sum_i P_{gq} \otimes (q_i + \bar{q}_i) + P_{gg} \otimes g \right).$$

In the expressions above, $q(x, Q^2)$ and $g(x, Q^2)$ denote the quark and gluon PDFs respectively, P_{ij} describes the $i \to j$ parton splitting function, which corresponds to the probability of the outgoing parton j to be emitted at a virtuality scale Q^2 and carry a fraction x/y of the mother parton momentum, and $\otimes$ is a symbol for the convolution integral:

$$P \otimes f \equiv \int_x^1 \frac{dy}{y} f_q(y) P \left(\frac{x}{y} \right).$$

As mentioned above, the PDFs have to be determined experimentally. This is done by several collaborations through fits (with typically 10 to 30 parameters) to experimental data. As an example, Fig. 4.2 shows the next-to-leading order (NLO) PDFs obtained by the ATLAS Collaboration at $Q^2 = 10$ GeV^2. Comparing this with PDFs at different Q^2 (e.g. in Ref. [6]), it can be observed that, with the increase of Q^2, the shape of the PDFs changes to favor lower x values. At low x values the gluon PDF is dominating (and this effect increases with Q^2), while for high x values the PDFs of the valence quarks are more relevant.

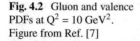

Fig. 4.2 Gluon and valence
PDFs at $Q^2 = 10$ GeV^2.
Figure from Ref. [7]

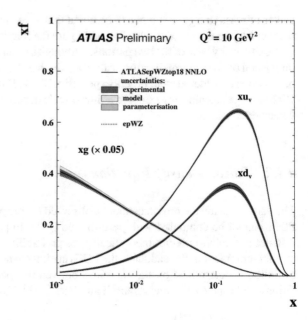

4.2 Event Simulation

The main steps in the simulation of a *pp* collision consist of:

- Computation of the matrix elements (MEs) for the hard process.
- Parton shower (PS) evolution and matching between PS and MEs.
- Hadronization and decay of unstable particles.
- Simulation of the underlying event and pileup.

In the next sections, each step is summarized in its main features.

4.2.1 Matrix Element

As discussed e.g. in Ref. [8], the partonic cross-section for the production of the
final state X starting from the partons i and j, necessary to compute Eq. 4.1, can be
expressed at all orders as:

$$\hat{\sigma}_{ij \to X} = \sum_{k=0}^{\infty} \int d\Phi_{X+k} | \sum_{l=0}^{\infty} \mathcal{M}_{X+k}^{(l)} |^2 . \tag{4.2}$$

In this expression, the index k denotes the number of final-state quarks and gluons
produced in addition to X (legs) and $\mathcal{M}_{X+k}^{(l)}$ the amplitude to produce the final state
$X + k$ computed with l virtual loops. These sums can not be computed to infinity,

and the perturbative order of the cross-section is identified by the number of extra partons and by the number of loops. In particular:

- The lowest possible order for the calculation of $\hat{\sigma}_{ij \to X}$ is the leading order (LO), where $k = l = 0$.
- $l = 0, k = n$ represents the LO computation for the production of $X + n$ jets.
- $k + l \leq n$ corresponds to a N^nLO prediction for the production of X, while N^{n-k}LO for the production of X in association with k jets.

At each order, the computation of the MEs implies a choice of the factorization and renormalization scales (μ_F and μ_R, respectively). These scales are not predetermined, and are typically set to values related to the characteristic energy scale of the considered physical process. The impact of this subjective choice is taken into account by evaluating each production cross-section at different scales (typically varying the nominal scale by a factor of two up and down), and assigning the difference as a systematic uncertainty in the cross-section estimate.

4.2.2 Parton Shower

A theorem by Kinoshita, Lee and Nauenberg (KLN theorem) [9, 10] guarantees that, when computing inclusive cross-sections, the logarithmic divergences arising from collinear splitting cancel against the virtual corrections order by order in perturbation theory. This does not hold anymore when we are interested in the computation of a differential cross-section, for example with a specific number of accompanying final-state extra partons, as it could be the case $l = 0, k = 1$ in Eq. 4.2. In this case, the kinematic of the basic inclusive event is simulated at fixed order, and the QCD emission process (splitting) is carried out by the PS algorithms [11], which generate an ordered sequence of emissions with decreasing angle or energy; the PS approximation consists in retaining only singular parts of the MEs, namely those corresponding to low angles and energies, obtaining a leading logarithm (LL) accuracy. Starting from the cross-section for the production of n particles, the cross-section for the production of $n + 1$ particles is calculated based on the parton splitting function P_{ij} (described in Sect. 4.1.2):

$$d\sigma_{n+1} \approx d\sigma_n \frac{\alpha_s}{2\pi} \frac{dt}{t} P_{ij}(z, t) \, dz \; .$$

The probability for a parton to evolve from an energy t to a lower energy t' without splitting is encoded in the Sudakov form factor:

$$\Delta_i(t, t') = \exp\left(-\sum_{j \in \{q, g\}} \int_t^{t'} \frac{d\bar{t}}{\bar{t}} \int_{z_{\min}}^{z_{\max}} dz \frac{\alpha_s}{2\pi} \frac{1}{2} P_{ij}(z, t)\right) , \qquad (4.3)$$

There are three types of emission processes that are described by the PS: $g \to q\bar{q}$, $g \to gg$ and $q \to qg$. Equation 4.3 is sampled with MC techniques to produce the sequence of splittings, until the hadronization scale is reached; this describes final-state radiation (FSR).

Also incoming particles can emit extra partons, giving rise to initial-state radiation (ISR). While in the case of FSR the first emissions are harder than subsequent ones, for ISR the ordering is inverted, and the showering is performed with a backwards-evolution algorithm [12].

4.2.3   Matching

When a process F is simulated with LO MEs plus PS, the hardest extra jet is described at LL accuracy. One might wish to improve the accuracy of the description of the event by adding the LO MEs for the process F with the addition of one extra parton (F+1). The naive addition of these two pieces leads to double counting of configurations. This double-counting problem worsens with the increase of the number of extra legs that we want to add to the MEs.

Different methods have been designed to match MEs to showers such that, for each order in perturbation theory, double-counting is avoided. The three main strategies are [13]:

Unitarity This approach consists in correcting the shower splitting functions by multiplicative factors obtained as the ratio of the ME to PS approximation:

$$\text{Matched} = \text{Approximate} \frac{\text{Exact}}{\text{Approximate}} .$$

When these correction factors are inserted in the shower evolution, they guarantee that the shower evolution of the process F describes correctly also the F+1 MEs, without actually adding the F+1 sample. This strategy has traditionally been worked out only for one extra parton emission.

Subtraction The subtraction approach consists in correcting the PS by the difference between the MEs and the PS:

$$\text{Matched} = \text{Approximate} + (\text{Exact} - \text{Approximate}) .$$

With this strategy, the corrections are not resummed and the events are weighted. This is the strategy used in MC@NLO [14–16], and the POWHEG method [17] is a hybrid between this and the unitarity approach.

Slicing The slicing approach divides the phase space into two regions: one described mainly by the MEs (by vetoing shower emissions below a cutoff scale, the matching scale), and one mainly by the PS:

$$\text{Matched(above matching scale)} \approx \text{Exact}(1 + \mathcal{O}(\alpha_s)) \, ,$$

$$\text{Matched(below matching scale)} = \text{Approximate} + (\text{Exact} - \text{Approximate})$$

$$\approx \text{Approximate} \, ,$$

where the last approximation holds because, below the matching scale, the MEs and the PS give similar results. With the slicing strategy, the corrections are not resummed and the events are weighted, but the weights are all positive. This approach can be extended beyond the first extra emission. Examples of this approach are the CKKW [18] and MLM [19, 20] prescriptions. The MLM approach at NLO is known as FxFx [21].

4.2.4 Hadronization

At the end of the shower we have emissions at very low energy, which correspond to high values of the strong coupling constant. When the strong coupling becomes of order unity, the interaction is very strong and this is presumably what causes the confinement of the colored partons into colorless hadrons (primary hadrons), that can in turn decay to secondary hadrons. This transition is described through phenomenological models since the hadronization scale, that corresponds also to the infrared (IR) cutoff of the parton shower, is outside the perturbative regime of QCD. At the end of the shower, the color, flavor and momentum distributions are already organized, and the hadronization process can only cause a local redistribution. The two main hadronization models used in MC generators are:

String Model   The string model [22] is based on linear confinement: the potential energy between two colored particles increases linearly with their distance, when the distance is greater than about 1 fm (while at shorter distances also a Coulomb term is present). The term "string model" comprises several different models, among which the most used nowadays is the Lund model [23, 24]. A color string forms that joins the final state in a string configuration of field energies, and hadrons originate from the breakups of the string. The proportionality constant of the linear inter-quark potential can be measured from quarkonia spectra or from lattice QCD, and results to be ≈ 1 GeV/fm. The main difference between quarks and gluons resides in the fact that, while quarks are connected to one single string, gluons are connected to two strings and have therefore a rate for hadron production twice that of quarks. The fragmentation function determines the probability of a given hadron to carry a certain fraction of the available momentum. In the Lund model the form of the fragmentation function is constrained by the left-right symmetry necessary to make the model independent of the sequence of string breakups. The resulting function depends on the mass and p_T of the hadron, leading to heavier hadrons carrying on average a higher fraction of the momentum.

Cluster Model   The cluster model is based on the preconfinement property of QCD. Color-singlet combinations of partons, referred to as clusters, are formed during the parton shower, and then decay to (possibly unstable) hadrons. Heavier clusters may first split through non-perturbative processes, and decay first to a pair of clusters or to a cluster and a hadron; this process continues until all clusters are transformed into hadrons. The mass distribution of the preconfined clusters results to be independent of the scale and nature of the original hard process. This universal distribution of the cluster mass peaks below 1 GeV, with a tail that extends to above 10 GeV.

These two hadronization models were developed in the 1980s, and since then there has been no fundamental progress in the theoretical understanding of hadronization. When the simulation of the events stops before hadronization, it is referred to as "parton-level".

4.2.5 Underlying Event

The partons from the colliding protons that do not participate in the hard scattering can undergo interactions as well, giving origin to multiple parton interactions in the same collision (underlying event). These secondary interactions are in general of lower momentum, since the ME is larger for low momentum transfer, and will contribute to the activity along the beam direction, less in the transverse plane. When the two protons pass through each other the likelihood of having multiple parton interactions depends on the overlap between the two protons. To model the underlying event, the assumption is made that these processes are $2 \to 2$ scattering; the ME for these diverges at small angle, so the modeling is dependent on the chosen p_T cutoff. Because of the low-p_T nature of the underlying event, its description is based on phenomenological models [25, 26]. To model the data, color reconnection between the primary interaction and the underlying event is needed.

4.2.6 Pileup

The term pileup refers to additional pp interactions taking place in the same bunch crossing (in-time pileup) or in events in different bunch crossings (out-of-time pileup). The presence of pileup challenges the reconstruction of the event, as it gives rise to extra activity overlapping with the products of the hard-scattering. The techniques used to model in-time pileup are the same as for the underlying event; for out-of-time pileup, similar methods are used but it is necessary to simulate the time response of the detector electronics to collisions from the previous bunch crossing.

4.3 Monte Carlo Generators

The simulations described in the above sections are carried out by event generators, that can be either general purpose, if they can reproduce all of the steps of the event generation, or specialized to one functionality. In this section the main characteristics of the generators used in the analyses described in this thesis are reviewed.

4.3.1 General Purpose Generators

Several independent software packages allow to study the effects of different modeling of the hadronization and different choices for the parton shower. The three main general-purpose event generators are:

PYTHIA   [27, 28] is a general purpose generator that uses LO MEs for $2 \rightarrow n$ ($n \leq 3$) processes; it is capable of simulating both hard and soft interactions, including ISR and FSR and multiple parton interactions. It uses a p_T-ordered parton shower and the Lund string hadronization model. It is commonly used as a PS generator, interfaced with a different generator that computes the MEs.

HERWIG   [29–31] has the same capabilities as PYTHIA with few small differences. It computes $2 \rightarrow 2$ LO MEs. The partons shower is ordered by the angle of the emitted parton. Gluon splitting processes ($g \rightarrow q\bar{q}$ and $g \rightarrow gg$) in the collinear approximation are not symmetric in the azimuthal direction due to interference of positive and negative helicity states in the original gluon. While PYTHIA uses a method that takes these affects into account only partially [32], HERWIG uses one that fully includes spin correlations [33]. The hadronization is based on the cluster model. HERWIG is typically interfaced with the standalone software JIMMY [34] that simulates the underlying event.

SHERPA   [35] The SHERPA event generator can provide multi-leg MEs both at LO (up to four extra partons) and at NLO (up to two extra partons). The matching between the ME and the dipole-type parton shower [36] follows the CKKW prescription. It uses a cluster hadronization model.

The recent versions of all three generators are coded in C++, but HERWIG and PYTHIA were originally developed in Fortran.

4.3.2 Matrix Elements Generators

The generators described in this section do not provide a full description of the event, but aim instead at improving the computation of the MEs, and can be afterwards interfaced to a general purpose generator to simulate parton shower, underlying event and pileup. The most common ME generators are:

Powheg- Box   [37] In this framework it is possible to implement NLO ME computations, using the 5-flavor scheme. It uses the Powheg method for matching.

MadGraph5_aMC@NLO   [38] This generator can compute MEs at LO for any user-
specified Lagrangian, and at NLO accuracy for selected processes. The NLO
calculation implements the MC@NLO method. It is then interfaced to a parton
shower using the MLM prescriptions at LO and the FxFx prescription at NLO.

4.3.3 Specialized Generators

The specialized generators provide a better description of one specific aspect of the
MC simulation. Some of them are:

EvtGen   [39] This package simulates the decay of heavy flavor particles, in particular B and D hadrons. In the simulation of the decay it uses decay amplitudes
instead of probabilities and it includes spin correlations. When interfaced with
other event generators, it can be used to re-decay the heavy flavor particles, substituting original decay chains by the more sophisticated simulation by EvtGen.

Tauola   [40] General purpose event generators treat tau leptons as stable particles. The tau decays are then handled by separated packages like Tauola, which
includes leptonic and semileptonic decays, paying attention to the tau polariza-
tion. Since the format of the tau-related information is generator-dependent and
the results of the original generator need to be replaces, also the input and output
formats of Tauola depend on the generator it is interfaced with.

Photos   [41] This generator handles electromagnetic radiation, estimating the size
of the quantum electrodynamics (QED) bremsstrahlung in the collinear approximation, and is used e.g. by Tauola.

4.4 Detector Simulation

The outcomes of the event simulation are the four-vectors of all the stable particles
produced in the final state of the event, stored in the standard HepMC format [42].
When this information is used directly, the analysis is referred to as "particle level"
analysis; furthermore, if we want to filter the simulated events based on the final
state, this can be done at this stage. While the event generation can already provide
information on the kinematic of the event, it is not enough to compare the MC
simulations with the data collected by the ATLAS detector. After the event simulation,
the ATLAS simulation chain [43] (described in Fig. 4.3) proceeds with the emulation
of the interaction of the particles with the detector and the signal generated in each
of the detector's subsystems, which is done with the Geant4 package [44]. The
configuration of the detector, including any misalignment, can be set at run time, and
the energy depositions are recorded as hits. The digitalization step takes the input

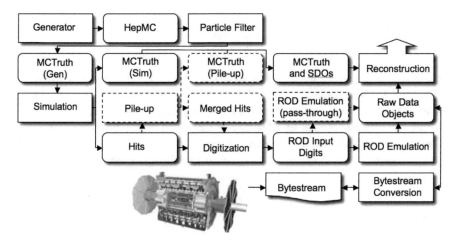

Fig. 4.3 The ATLAS simulation chain, compared with the processing of the recorded data; square-cornered boxes represent algorithms, while data objects are represented as rounded boxes. Figure from Ref. [43]

hits from the hard scattering, underlying event and pileup and transforms them into detector signals, adding also detector noise. Also the response of the Level 1 (L1) trigger (described in Sect. 3.3.7) is simulated at this stage. The final output of the digitalization is the raw data object (RDO) file, to which the output of the ATLAS detector itself, which is in "bytestream" format, can be converted as well. The HLT decision and the event reconstruction run on the RDO data format. After the detector simulation, the MC simulated events are treated in the same manner as data, going through the object reconstruction procedures described in Chap. 5.

The full simulation of the interaction of particles with the ATLAS detector is a CPU-intensive task. ATLFAST-II [43] is a fast simulation method making use of a simplified detector description. It has two components: the fast ATLAS tracking simulation (Fatras) [45], to emulate the response of the Inner Detector (ID) and of the muon system, and the fast calorimeter simulation (FastCaloSim) [46], that takes care of the simulation of the calorimeters. The default ATLFAST-II simulation uses the full GEANT4 simulation for the ID and the muon spectrometer, while it uses FastCaloSim to emulate the energy deposited in the calorimeters using a parametrization of the longitudinal and lateral energy profile. ATLFAST-IIF uses both FastCaloSim for the calorimeters and Fatras for the tracking systems. The output of ATLFAST-II includes all the properties necessary to run the same event reconstruction as with GEANT4 or the real data.

The simulated event samples are normalized to the highest-available-order cross-section, and events are reweighted so that the simulated pileup distribution matches that observed in the data. Despite the accurate simulation, residual differences can be present in the reconstruction and selection efficiency in data and MC simulation. The simulated reconstruction and selection efficiencies are corrected with multiplicative scale factors (SFs), defined as:

$$SF = \frac{\epsilon_{data}}{\epsilon_{MC}} \, ,$$

where ϵ_{data} and ϵ_{MC} are measured in dedicated data calibration samples and in the equivalent MC simulation, respectively. These SFs can be function of the kinematic of the physics objects in the event (often p_T and η). Some examples of SFs for the physics objects relevant for the analyses described in this thesis are provided in Chap. 5. Analogously, energy scale and resolution of the different physics objects in the simulation are corrected to match the corresponding measurements in data.

References

1. Butterworth JM, Dissertori G, Salam GP (2012) Hard processes in proton-proton collisions at the large hadron collider. Ann Rev Nucl Part Sci 62:387. arXiv:1202.0583 [hep-ex]
2. Collins JC, Soper DE (1987) The theorems of perturbative QCD. Annu Rev Nucl Part Sci 37:383
3. Dokshitzer YL (1977) Calculation of the structure functions for deep inelastic scattering and e+ e- annihilation by perturbation theory in quantum chromodynamics. Sov Phys JETP 46:641. [Zh Eksp Teor Fiz 73:1216 (1977)]
4. Gribov VN, Lipatov LN (1972) Deep inelastic e p scattering in perturbation theory. Sov J Nucl Phys 15:438. [Yad Fiz 15:781 (1972)]
5. Altarelli G, Parisi G (1977) Asymptotic freedom in parton language. Nucl Phys B 126:298
6. Martin AD, Stirling WJ, Thorne RS, Watt G (2009) Parton distributions for the LHC. Eur Phys J C 63:189. arXiv:0901.0002 [hep-ph]
7. ATLAS Collaboration (2018) Determination of the parton distribution functions of the proton from ATLAS measurements of differential W and Z/γ^* and $t\bar{t}$ cross sections. ATL-PHYS-PUB-2018-017. https://cds.cern.ch/record/2633819
8. Skands PZ (2011) QCD for collider physics. In: Proceedings, high-energy physics. Proceedings, 18th European School (ESHEP 2010): Raseborg, Finland, June 20–July 3, 2010. arXiv:1104.2863 [hep-ph]
9. Kinoshita T (1962) Mass singularities of Feynman amplitudes. J Math Phys 3:650
10. Lee TD, Nauenberg M (1964) Degenerate systems and mass singularities. Phys Rev 133:B1549
11. Fox GC, Wolfram S (1980) A model for parton showers in QCD. Nucl Phys B 168:285
12. Sjostrand T (1985) A model for initial state parton showers. Phys Lett B 157:321
13. Giele WT, Kosower DA, Skands PZ (2011) Higher-order corrections to timelike jets. Phys Rev D 84:054003 arXiv:1102.2126 [hep-ph]
14. Frixione S, Webber BR (2002) Matching NLO QCD computations and parton shower simulations. JHEP 06:029. arXiv:hep-ph/0204244 [hep-ph]
15. Frixione S, Nason P, Webber BR (2003) Matching NLO QCD and parton showers in heavy flavor production. JHEP 08:007. arXiv:hep-ph/0305252 [hep-ph]
16. Frixione S, Webber BR. The MC and NLO 3.4 event generator. arXiv:0812.0770 [hep-ph]
17. Frixione S, Nason P, Oleari C (2007) Matching NLO QCD computations with parton shower simulations: the POWHEG method. JHEP 11:070. arXiv:0709.2092 [hep-ph]
18. Catani S, Krauss F, Kuhn R, Webber BR (2001) QCD matrix elements + parton showers. JHEP 11:063. arXiv:hep-ph/0109231 [hep-ph]
19. Mangano ML, Moretti M, Piccinini F, Treccani M (2007) Matching matrix elements and shower evolution for top-quark production in hadronic collisions. JHEP 01:013. arXiv:hep-ph/0611129 [hep-ph]

20. Mrenna S, Richardson P (2004) Matching matrix elements and parton showers with HERWIG and PYTHIA. JHEP 05:040. arXiv:hep-ph/0312274 [hep-ph]
21. Frederix R, Frixione S (2012) Merging meets matching in MC@NLO. JHEP 12:061. arXiv:1209.6215 [hep-ph]
22. Artru X, Mennessier G (1974) String model and multiproduction. Nucl Phys B 70:93
23. Andersson B, Gustafson G, Ingelman G, Sjostrand T (1983) Parton fragmentation and string dynamics. Phys Rept 97:31
24. Andersson B (1997) The Lund model. Camb Monogr Part Phys Nucl Phys Cosmol 7:1
25. ATLAS Collaboration (2014) ATLAS Pythia 8 tunes to 7 TeV data. ATL-PHYS-PUB-2014-021. https://cds.cern.ch/record/1966419
26. Skands PZ (2010) Tuning Monte Carlo generators: the perugia tunes. Phys Rev D 82:074018. arXiv:1005.3457 [hep-ph]
27. Sjostrand T, Mrenna S, Skands PZ (2006) PYTHIA 6.4 physics and manual. JHEP 05:026. arXiv:hep-ph/0603175 [hep-ph]
28. Sjöstrand T, Ask S, Christiansen JR, Corke R, Desai N, Ilten P, Mrenna S, Prestel S, Rasmussen CO, Skands PZ (2015) An introduction to PYTHIA 8.2. Comput Phys Commun 191:159. arXiv:1410.3012 [hep-ph]
29. Corcella G, Knowles IG, Marchesini G, Moretti S, Odagiri K, Richardson P, Seymour MH, Webber BR (2001) HERWIG 6: an event generator for hadron emission reactions with interfering gluons (including supersymmetric processes). JHEP 01:010. arXiv:hep-ph/0011363 [hep-ph]
30. Bahr M et al (2008) Herwig++ physics and manual. Eur Phys J C 58:639. arXiv:0803.0883 [hep-ph]
31. Bellm J et al (2016) Herwig 7.0/Herwig++ 3.0 release note. Eur Phys J C 76:196. arXiv:1512.01178 [hep-ph]
32. Webber BR (1987) Quantum correlations in QCD Jets. Phys Lett B 193:91
33. Collins JC (1988) Spin correlations in Monte Carlo event generators. Nucl Phys B 304:794
34. Butterworth JM, Forshaw JR, Seymour MH (1996) Multiparton interactions in photoproduction at HERA. Z Phys C 72:637. arXiv:hep-ph/9601371 [hep-ph]
35. Gleisberg T, Hoeche S, Krauss F, Schonherr M, Schumann S, Siegert F, Winter J (2009) Event generation with SHERPA 1.1. JHEP 02:007. arXiv:0811.4622 [hep-ph]
36. Schumann S, Krauss F (2008) A Parton shower algorithm based on Catani-Seymour dipole factorisation. JHEP 03:038. arXiv:0709.1027 [hep-ph]
37. Alioli S, Nason P, Oleari C, Re E (2010) A general framework for implementing NLO calculations in shower Monte Carlo programs: the POWHEG BOX. JHEP 06:043. arXiv:1002.2581 [hep-ph]
38. Alwall J, Frederix R, Frixione S, Hirschi V, Maltoni F, Mattelaer O, Shao HS, Stelzer T, Torrielli P, Zaro M (2014) The automated computation of tree-level and next-to-leading order differential cross sections, and their matching to parton shower simulations. JHEP 07:079. arXiv:1405.0301 [hep-ph]
39. Lange DJ (2001) The EvtGen particle decay simulation package. Nucl Instrum Meth A 462:152
40. Jadach S, Kuhn JH, Was Z (1990) TAUOLA: a library of Monte Carlo programs to simulate decays of polarized tau leptons. Comput Phys Commun 64:275
41. Barberio E, van Eijk B, Was Z (1991) PHOTOS: a universal Monte Carlo for QED radiative corrections in decays. Comput Phys Commun 66:115
42. Dobbs M, Hansen JB (2001) The HepMC C++ Monte Carlo event record for high energy physics. Comput Phys Commun 134:41
43. ATLAS Collaboration (2010) The ATLAS simulation infrastructure. Eur Phys J C 70:823. arXiv:1005.4568 [physics.ins-det]
44. GEANT4 Collaboration, Agostinelli S et al (2003) GEANT4: a simulation toolkit. Nucl Instrum Meth A 506:250
45. Edmonds K, Fleischmann S, Lenz T, Magass C, Mechnich J, Salzburger A. The fast ATLAS track simulation (FATRAS). https://cds.cern.ch/record/1091969
46. ATLAS Collaboration (2010) The simulation principle and performance of the ATLAS fast calorimeter simulation FastCaloSim. ATL-PHYS-PUB-2010-013. https://cds.cern.ch/record/1300517

Chapter 5
Event Reconstruction

The particles produced in the *pp* collisions in the center of the ATLAS detector interact with the detector material as discussed in Chap. 3. As a result of these interactions, electrical signals are recorded. Event reconstruction is the process of recombining these digital signals and interpreting them as tracks and energy deposits in the calorimeters. Finally, a particle identification step is performed, where the information from the relevant subdetectors is combined to reconstruct as accurately as possible a candidate physics object. This chapter describes the reconstruction and identification of the objects used in the analyses discussed in this thesis: tracks and vertices, hadronic jets, muons, electrons and missing transverse momentum.

5.1 Tracks and Primary Vertices

In ATLAS the identification of tracks from charged particles relies on the information collected by the ID. The tracking information is crucial to the reconstruction and identification of many types of particles, including electrons, muons, and the jets originating from the hadronization of a *b*-quark. Charged particles traversing the ID deposit energy through ionization, which is read out as hits; in the Pixel detector each hit corresponds to a space point, while in the SCT the space points are obtained as pairs of hits from each side of the modules. The space points are used to reconstruct the trajectory of the charged particles, which is helicoidal and with radius inversely proportional to their momentum, since the ID is surrounded by a solenoidal magnetic field. The precision on the position measurement of the track depends on the granularity of the different subsystems of the ID.

After the point of closest approach (perigee) to a given reference is defined, the trajectory of the track can be described by five parameters:

$$\theta, \ \phi, \ q/p, \ d_0, \ z_0 \,,$$

© Springer Nature Switzerland AG 2020
C. Rizzi, *Searches for Supersymmetric Particles in Final States
with Multiple Top and Bottom Quarks with the Atlas Detector*, Springer Theses,
https://doi.org/10.1007/978-3-030-52877-5_5

where θ and ϕ are the azimuthal and polar angle, q/p is the ratio of the charge of the track to the track momentum, and d_0 and z_0 are the distance to the point of closest approach to the vertex in the transverse plane and along the z-axis.

Primary tracks, originating from charged particles with a life time longer that 3×10^{-11} s produced directly in the hard-scattering vertex, are reconstructed with an inside-out approach [1]: the seed of the reconstruction are three hits in the silicon detector, and then compatible hits in the outer layers of the ID are added with a Kalman Filter [2, 3]. The TRT segments that are not associated with primary tracks are used as starting point to reconstruct tracks from long-lived particles or from material interaction, with a back-tracking that extrapolates the TRT information to the pixel hits.

Random groups of hits can be wrongly reconstructed as belonging to the helical trajectory of a track (fake tracks). The amount of fake tracks increases with the increase of pileup, and can be reduced by tightening the selection criteria of the track, at the expense of reconstruction efficiency. Three different selection criteria are used for the data collected in 2015 and 2016 (Loose, Loose-Primary and Tight-Primary), that differ in the requirements on the hits and holes (elements where a hit was expected but was not registered) in the different ID layers. The track reconstruction efficiency is measured in MC simulations as the ratio of the reconstructed tracks matched to a generated charged particle over the total number of generated charged particles. The reconstruction efficiency as a function of the track η and p_T is shown in Fig. 5.1 for Loose and Tight-Primary tracks.

Tracks are the starting point for the identification of the interaction points, referred to as primary vertexes (PVs). PVs are reconstructed through a vertex finding algorithm [5], and then the vertex fitting algorithm identifies the vertex position and refits the tracks adding the constraint of the reconstructed interaction point. The LHC operates in a high-luminosity regime, which makes it likely to have multiple pp interactions per bunch crossing, and therefore multiple reconstructed PV candidates. Once all the PV candidates are reconstructed, the one with the highest sum of the squared

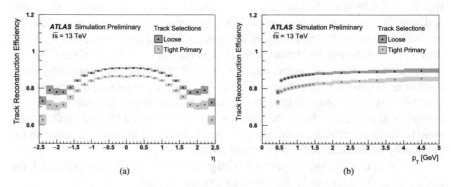

Fig. 5.1 Track reconstruction efficiency, evaluated by using minimum bias simulated events, as a function of truth η **a** and p_T **b** for Loose and Tight Primary track selections. The bands indicate the total systematic uncertainty. Figure from Ref. [4]

transverse momenta of its associated tracks ($\sum_i^{N-tracks} p_{T,i}^2$) is identified as the hard scattering PV, and the position of physics objects is recomputed with respect to its coordinates. The other vertices are named pileup vertices, and their number is correlated to the number of interactions per bunch crossing.

5.2 Jets

Because of confinement, quarks and gluons produced in the collisions give origin to a collimated spray of hadrons (jets) that move in the direction of the original parton. When jets interact with the detector, they loose most of their energy as deposits in the calorimeter systems, which are then grouped together aiming at reconstructing the characteristics of the original parton.

5.2.1 Clusters

The first step in the jet reconstruction is the procedure that groups the calorimeter cells in three dimensional objects referred to as topological clusters (topoclusters) [6, 7]. Topoclusters are built starting from seed cells with a signal-to-noise ratio higher than 4. All the neighboring cells with signal-to-noise ratio higher than two are added with an iterative procedure, and finally a ring of guard cells are added independently of their signal. Topoclusters are calibrated at the electromagnetic (EM) scale, which means that the proportionality constant between the readout current and the particle energy is correct only for particles of an EM shower.

5.2.2 Jet-Finding Algorithms

The topoclusters are then grouped together by a jet-finding algorithm. Different algorithms are available, and in particular the algorithms of the k_T-family merge clusters according to the metric $d_{i,j}$, defined as:

$$d_{i,j} = min\left(k_{T,i}^{2n}, k_{T,i}^{2n}\right) \frac{\Delta R_{i,j}^2}{R^2}, \tag{5.1}$$

where $k_{T,i}$ is the transverse momentum of the cluster, $\Delta R_{i,j}$ is the angular distance defined as in Eq. 3.4, R is a fixed parameter, whose value sets the size of the jet, and n is the parameters that defines the kind of algorithm we are using and therefore the shape of the resulting jets. Equation 5.1 defines the distance between two clusters, while the cluster-beam distance is defined as:

$$d_{i,B} = k_{T,i}^{2n} \; .$$

The grouping of clusters follows an iterative approach:

1. For each topocluster, the distances $d_{i,j}$ and $d_{i,B}$ are calculated.
2. If, for some i and j, $d_{i,j} < d_{i,B}$, the two clusters with the smallest $d_{i,j}$ are grouped.
3. Otherwise, if $d_{i,B} < d_{i,j}\ \forall i \neq j$ the i-th cluster is defined as a jet.
4. This procedure is iterated until all inputs have been classified into jets.

Depending on the value of the parameter n, we can distinguish different algorithms:

- $n = 0$: Cambridge-Aachen. The grouping of the clusters depends only on geometrical considerations and not on their momentum.
- $n = 1$: k_T algorithm. Soft clusters are grouped first.
- $n = -1$: anti-k_T algorithm. Groups hard objects first; the shape of the jets is more regular than in the two previous cases and is a cone of radius R.

The choice of a particular algorithm results in different shapes of the jets, as discussed in Ref. [8]. The standard algorithm used by ATLAS is the anti-k_T algorithm, which leads to jets with a more regular shape in the (η-ϕ) plane.

5.2.3 Jet Calibration

As mentioned previously, the inputs to the jet-finding algorithm are calibrated at the EM scale, and its coordinates refer to the center of the detector. To access a more precise measurement of the jet energy and kinematics, a sequence of calibration steps is applied; the standard ATLAS corrections are [9]:

Origin correction   The direction of the jet is changed to point to the reconstructed hard-scattering PV rather than to the centered of the detector. This correction improves the η resolution of the jets.

Pileup correction   Multiple collisions in the same bunch crossing (in-time pileup), as well as residual energy from previous collisions (out-of-time pileup), affect the jet energy reconstruction. The effect of pileup is corrected in two steps [10, 11]: a first correction, dependent on the number of PVs, uses the jet area to subtract form the jet energy the average energy form pileup events. The jet area is measured with ghost-association: simulated ghost particles of infinitesimal momentum are added to the event uniformly in solid angle prior to jet reconstruction, and the jet area is computed from the fraction of ghost particles associated to the jet after the clusters are merged. A second correction based on the number of PVs and on the number of interactions per bunch crossing is then applied to disentangle the effect of in-time pileup and out-of-time pileup.

Absolute calibration   The absolute jet energy scale (JES) and η correction is derived comparing in MC the truth energy of a jet (defined as the energy of the

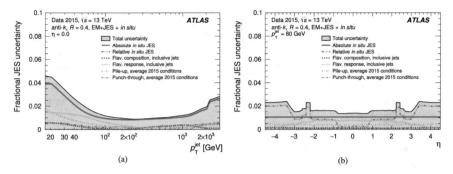

Fig. 5.2 Combined uncertainty in the JES of fully calibrated jets as a function of **a** jet p_T at $\eta = 0$ and **b** η at $p_T = 80$ GeV. Figure from Ref. [9]

truth jet with $\Delta R < 0.3$ from the calorimeter jet) with the reconstructed energy, and it also corrects for biases in the η reconstruction [12].

Global sequential calibration The global sequential calibration (GSC) [13] improves the JES resolution by deriving additional corrections based on individual jet properties, e.g. the number of associated tracks and the fraction of energy deposited in the various layers of the calorimeter.

In-situ calibration As a last stage, the data-driven calibration (in-situ) [14] corrects for the differences between MC simulation and data (arising e.g. from imperfect simulation of the detector response and material interaction). These corrections are derived from events where the jet p_T is balanced against other well measured objects. In the η-intercalibration, dijet events are used to correct the response of forward jets (with $0.8 < |\eta| < 4.5$) using well-measured central jets (with $|\eta| < 0.8$). The response of central jets is instead measured in Z+jets, γ+jets and multijet events; in Z/γ+jets events, the p_T of the jet is measured against the p_T of a well measured Z boson or photon, while multijet events are used to calibrate central high-p_T jets ($300 < p_T < 2000$ GeV) against well calibrated low-p_T central jets.

5.2.3.1   Jet Calibration Uncertainties

The calibration procedure described above implies a set of uncertainties. In particular, the ATLAS JES Run-2 calibration includes a set of 80 systematic uncertainties; 67 of those derive from the in-situ calibration [9], accounting for modeling uncertainties, sample statistical uncertainty and uncertainties in the calibration of other physics objects used in deriving the calibration. The other 13 systematic uncertainties derive from the pileup correction, the η-intercalibration and differences in the jet response and composition for jets of different flavors. The full combination of the uncertainties derived from the first 3.2 fb^1 of the Run 2 data is shown in Fig. 5.2 as a function of p_T at $\eta = 0$ and as a function of η at $p_T = 80$ GeV.

To allow an easier usage of the uncertainties in physics analyses, a reduced set of 19 nuisance parameterss (NPs) is provided: the 67 NPs from the in-situ calibration are reduced to six by keeping the five uncertainties of largest magnitude, plus a sixth one which is the sum in quadrature of the remaining ones. A further reduction is in place to group the remaining NPs into three components, and the NPs within a single component are added in quadrature; this leads to a correlation loss, whose effect is analysis dependent.

Jets with the same true energy can have different reconstructed energies; the distribution of the difference between true and reconstructed energy of a jet is modeled with a Gaussian, whose width is defined as jet energy resolution (JER). The measurements for the in-situ calibrations can be used also to access the differences in JER between data and MC [14, 15], and result into an additional NP.

5.2.4 Jet Vertex Tagger

Beside affecting the jet energy, high levels of pileup can also lead to the reconstruction of spurious jets not originating from the hard-scattering interaction. In ATLAS track-based algorithms are used for the identification of pileup jets [16, 17]. Tracks are matched to calorimeter jets by ghost-association [18]. Once the hard-scattering PV (PV_0) is identified, for each jet it is possible to compute the jet vertex fraction (JVF), the ratio of the scalar sum of the p_T of the tracks associated to the jet and originating from PV_0 to that of all the associated tracks:

$$JVF = \frac{\sum_m p_{T,m}^{track}(PV_0)}{\sum_n \sum_l p_{T,l}^{track}(PV_n)} .$$

(5.2)

In this definition the index n runs over all the vertices of the event. In high-pileup conditions the denominator in Eq. 5.2 increases; to correct for this dependence, the corrected jet vertex fraction (corrJVF) is introduced. This is a modified version of JVF, that takes into account the dependence on the number of pileup tracks (n_{track}^{Pu}):

$$corrJVF = \frac{\sum_m p_{T,m}^{track}(PV_0)}{\sum_l p_{T,l}^{track}(PV_0) + \frac{\sum_{n\geq 1} \sum_l p_{T,l}^{track}(PV_n)}{(k \cdot n_{track}^{Pu})}} ,$$

with $k = 0.001$. Another variable used to discriminate hard-scattering jets against pileup jets is the ratio of the scalar sum of the p_T of the tracks originating from PV_0 to the jet transverse momentum:

$$R_{pT} = \frac{\sum_k p_{T,k}^{track}(PV_0)}{p_T^{jet}} .$$

(5.3)

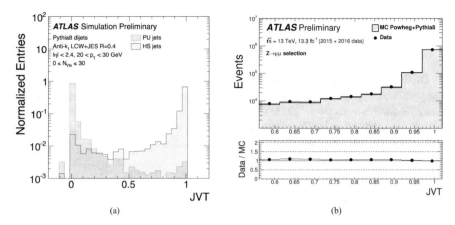

Fig. 5.3 **a** Distribution of JVT for pileup and hard-scatter jets with $20 < p_T < 30$ GeV. Figure from Ref. [17]. **b** The JVT distribution, in Powheg+Pythia8 MC and in 2015+2016 data, of jets balanced against Z bosons decaying to muons. Figure from Ref. [19]

The corrJVF and R_{pT} variables are combined in a two-dimensional likelihood that constitutes a single tagger, the jet vertex tagger (JVT). The distribution of JVT for hard-scattering and pileup jets in MC simulation is shown in Fig. 5.3a, while Fig. 5.3b shows the comparison of the JVT distribution in data and MC in a dimuon selection.

5.2.5 Jet Cleaning

Beside pileup jets, other spurious jets come from the non-collision background; this type of background includes muons originating from secondary cascades from beam losses, in which case we speak of beam-induced background, and from cosmic rays. These muons leave energy deposits in the calorimeters while traversing the detector, which can be interpreted as jets. Also coherent noise from the calorimeters can give rise to fake jets. In ATLAS a set of quality criteria are designed to reject jets not originating from pp collisions [20]. These quality criteria are rely on variables based on:

- Ionization signal shape in the LAr calorimeters, to remove mainly fake jets from calorimeter noise.
- Ratios of energies, e.g. the ratio of the energy deposited in the electromagnetic calorimeter to the total energy, or the ratio of energy in different calorimeter layers, that can be used to discriminate against jets from beam-induced background or calorimeter noise.

- Tracks associated with the jets, and in particular variables similar to R_{pT} defined in Eq. 5.3, that have in general lower value for fake jets than for jets originating from pp collisions.

Different thresholds for the selections on these variable distinguish the two working points, *BadLoose* and *BadTight*, which have an efficiency respectively of 99.5% and 95% for jets with $p_T > 20$ GeV, while for jets with $p_T > 100$ GeV the efficiency of the two working points increases to 99.9% and 99.5%. The operating point (OP) used in the searches discussed in Chaps. 8 and 9 is *BadLoose*.

5.2.6 Re-clustered Jets

The angular separation between the decay products of a particle with mass m and transverse momentum p_T scales as:

$$\Delta R \approx \frac{2m}{p_T} .$$

This indicates that the ideal value of the R parameter described in Sect. 5.2.2 can vary depending on the event topology that we want to capture. For example, the decay products of a heavy particle with a transverse momentum much larger than its rest mass (boosted object) could be better described by a single jet with a larger R than with multiple jets with the "standard" 0.4 radius, as it happens e.g. in the decay of very energetic top quarks, W, Z or Higgs bosons produced at the LHC. Each different value of the R parameter requires a dedicated calibration following the steps described in Sect. 5.2.3. Therefore, it is not always possible to choose the optimal value of the jet radius. A possible solution to this problem comes from noticing that the same jet-finding algorithms used to group topoclusters can have different types of inputs. In particular, jets themselves can be used as input and grouped together, and in this case we speak of re-clustered jets [21]. Re-clustered jets are automatically calibrated as long as the input jets are, and also the jet uncertainties can be propagated directly. A comparison of the jet clustering obtained with anti-k_T $R = 1.0$ and by re-clustering anti-k_T $R = 0.3$ jets into anti-k_T $R = 1.0$ jets is shown in Fig. 5.4. It is possible to see how the jet axis is similar between the two cases. Re-clustered jets can be trimmed by removing the constituent small-R jets that have a p_T smaller than a defined fraction of the p_T of the original reclustered jet.

5.3 Jets from B-Hadrons

Jets originating from the hadronization of a b-quark (b-jets) can be identified thanks to the lifetime of B-hadrons (about 10^{-12} s), which is shorter than the typical lifetime

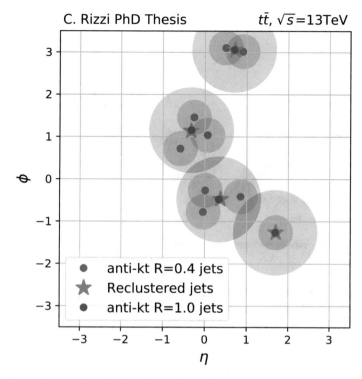

Fig. 5.4 Example event where jets have been clustered with anti-k_T with $R = 0.4$ (red dots) and anti-k_T $R = 1.0$ (blue dots), and re-clustered starting from anti-k_T $R = 0.4$ jets (green stars); circles of radius 0.4 and 1.0 are drawn centred around anti-k_T jets with $R = 0.4$ and $R = 1.0$ respectively

of hadrons containing only light quarks, but still long enough to allow the B-hadrons to travel distances of the order of the mm before decaying. A schematic view of the topology originating from a jet containing a B-hadron is shown in Fig. 5.5.

The procedure of identifying b-jets is referred to as b-tagging, and in ATLAS is performed using as input the tracks associated to the jets. As already mentioned in Sect. 3.3.3.1, between Run 1 and Run 2 a fourth pixel layer, the IBL, was added to the ATLAS detector, allowing a better impact parameter resolution and therefore improving substantially the b-tagging performance.

There are three families of b-tagging algorithms that can be combined through multivariate techniques. The basic algorithms can be based on:

Impact Parameter The transverse impact parameter of a track (d_0) is the point of closest approach to the PV in the transverse plane, while the longitudinal impact parameter ($z_0 \sin\theta$) is defined as the distance along the z-axis between the PV and the point of closest approach in the transverse plane. Because of the typical lifetime of B-hadrons, on average b-jets contain tracks with higher impact parameter than light-jets. The sign of the impact parameter is positive if the track extrapolation crosses the jet direction in front of the primary vertex,

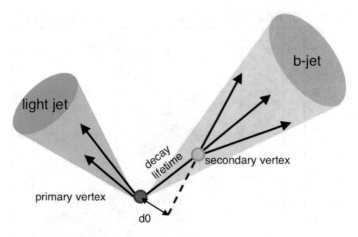

Fig. 5.5 Schematic view of the topology of a b-jet

and negative otherwise. The negative side of the impact-parameter distribution derives from the impact-parameter resolution, and can be used to calibrate light-jets. In ATLAS, two taggers make use of the information on the impact parameter [22]: IP2D, which is based on the significance of the transverse impact parameter (d_0/σ_{d_0}), and IP3D, which builds a two-dimensional template including also the significance of the longitudinal impact parameter ($z_0\sin\theta/\sigma_{z_0\sin\theta}$). The probability density function (PDF) for each flavor hypothesis (b, c, and light) is derived from MC simulation on a per-track basis, and then a log-likelihood ratio (LLR) of the different probabilities is computed, including the contribution from all tracks associated to the jet. For example, the LLR discriminating b-jets from light-jets is of the form $\sum_{i=1}^{N} \log \frac{p_b}{p_{light}}$, where the index N runs on all the tracks associated to the jet.

Secondary Vertex Finding   The SV1 algorithm [22] explicitly looks for a secondary vertex within a jet. All the track pairs in the jet are tested for a two-track vertex hypothesis, removing the pairs that are likely to originate from long-lived particles (e.g. K_0, Λ), photon conversion or hadronic interaction with the detector material. If a two-track vertex remains, a new single vertex is fitted with the tracks passing this selection. B-tagged jets are identified by high values of a likelihood discriminant, built using several variables including the decay-length significance, the invariant mass of all tracks associated with the vertex, the ratio of the sum of the energies of the tracks in the vertex to the sum of the energies of all tracks in the jet, and the number of two-track vertices.

Identification of the Decay Chain   The B-hadrons inside b-jets decay with an electroweak interaction, through which a b-quark decays preferentially to a c-quark, since the CKM matrix element $|V_{cb}|^2$ is much larger than $|V_{ub}|^2$. Hadrons containing a c-quark (D-hadrons) subsequently decay as well, giving rise to a topology with two decay vertices. While the resolution is often not enough to reconstruct the two vertices individually, the JetFitter algorithm [23] operates

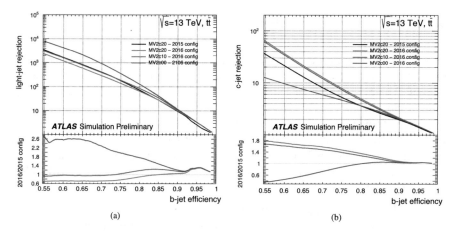

Fig. 5.6 Light-jet (**a**) and c-jet (**b**) rejection as a function of the b-jet efficiency for the MV2 algorithms. Figures from Ref. [25]

Table 5.1 Operating points for the MV2c10 b-tagging algorithm. The efficiency and rejection rates are computed for jets with $p_T > 20$ GeV from $t\bar{t}$ events. Table from Ref. [25]

BDT Cut Value	b-jet Efficiency (%)	c-jet Rejection	Light-jet Rejection	τ-Rejection
0.9349	60	34	1538	184
0.8244	70	12	381	55
0.6459	77	6	134	22
0.1758	85	3.1	33	8.2

assuming that they both lie on the same line, the flight axis of the B-hadron. The information on the event topology derived with JetFitter is then used in a likelihood function, from which three different templates (one for each flavor) are derived.

The default b-tagging algorithm used by ATLAS in the analysis of the 2015–2016 dataset is MV2c10 [24, 25], a multivariate algorithm based on a boosted decision tree (BDT) that combines the algorithms described above. MV2c10 belongs to the family of MV2 algorithms, which are trained on a $t\bar{t}$ sample using b-jets as signal and c-jets and light-jets as background, and differ in the relative fraction of c-jets and light-jets that are used in the training; in the case of MV2c10, the background sample in the training contains 15% of c-jets. Figure 5.6 shows the light-jet and c-jet rejection as a function of b-jet efficiency for different MV2 algorithms.

OPs are defined by a selection on the value of the BDT output, and are designed to have a specific b-jet efficiency. Table 5.1 shows the OPs defined for the MV2c10 algorithm.

5.3.1   B-Tagging Calibration and Uncertainties

The b-tagging efficiency can be different in MC simulation and data. The b-tagging efficiency, c-tagging efficiency and light mistag rate are measured in data for the OPs of Table 5.1, and the MC simulation is corrected with the SFs derived as the ratio of the efficiency in data and in MC. The SFs are derived on a per-jet basis in a parametric form based on jet p_T, η and truth flavor. For each MC simulated event, an event-level SF is derived by multiplying all the efficiency SFs for the b-tagged jets and all the inefficiency SFs for the jets that are not b-tagged. Several techniques are used in ATLAS to measure the b-tagging efficiency for the different jet flavors [26]. The calibrations used in the analyses described in this thesis are:

b-jets   The default b-tagging calibration for b-jets is based on a $t\bar{t}$ dileptonic sample. Events with exactly two opposite-sign leptons and two or three jets are selected, and a per-event likelihood is built containing the b-tagging weight PDF for a jet of a given flavor; the PDF for light-jets and c-jets is taken from MC, while the PDF for b-jets is the information that we want to extract from data. This last PDF is described by a histogram with only two bins, one below and one above the threshold to b-tag a jet.

c-jets   The analysis described in Chap. 8 uses a c-jet calibration based on events where a W boson is produced in association with a c-quark. The events selected are the ones where the W boson decays to an electron and a neutrino, and the D-hadron originating from the fragmentation of the c-quark decays to a muon. In $W + c$ production the electron and the muon in the final state have opposite charge, while most of the background processes have an equal number of same-sign and opposite-sign events. The number of $W + c$ events can therefore be obtained as the difference of these two categories. For the higgsino search described in Chapter 9, which was developed at a later time, a new calibration for c-jets, based on $t\bar{t}$ events [27], was available. This calibration selects $t\bar{t}$ events where one of the W bosons decays leptonically and the other one decays to a c-quark and an s-quark.

light-jets   The b-tagging efficiency of light-jets (mistag rate) is measured on an inclusive sample of jets, using the negative tag method [28]. The two main reasons that lead to b-tagged light-jets are the finite resolution of the impact parameter and the secondary vertices caused by long-lived particles and material interactions. If we consider only the first type of mistags, the signed impact parameter distribution will be symmetric around zero. The negative tag method is based on a modified version of the b-tagging algorithms, that takes as input impact parameters and decay lengths with reversed sign. The mistag rate is measured as the negative-tag efficiency of the jet sample, with MC-based correction factors that take into account the negative-tag rate for b- and c-jets and the effect of long-lived particles and material interactions.

The b-tagging scale factors are affected by multiple sources of uncertainty, which are reflected in uncertainties in the SFs. As an example, Fig. 5.7 shows the b-tagging SF for c-jets derived with the $t\bar{t}$ calibration 77% OP, and Fig. 5.8 shows the SFs for light-jets derived with the negative tag calibration for the same OP.

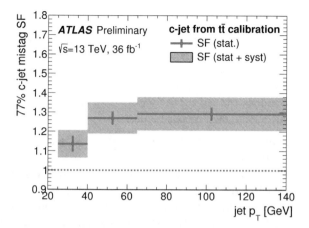

Fig. 5.7 *B*-tagging SF for *c*-jets for the OP corresponding to the 77% OP. Figure from Ref. [27]

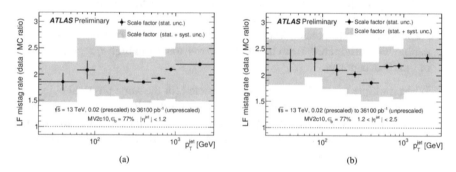

Fig. 5.8 *B*-tagging SF for light-jets for the OP corresponding to the 77% OP for **a** |η| < 1.2 and **b** 1.2 < |η| < 2.5. Figures from Ref. [28]

5.4 Muons

Muon reconstruction and identification [29] is based on the information collected by the ID, where muons are reconstructed as charged tracks, and by the MS.

5.4.1 Muon Reconstruction

Both the ID and the MS perform the reconstruction of muon candidates independently and then the information is combined to build the muon candidates used in physics analyses. The reconstruction of muon tracks in the ID proceeds as described in Sect. 5.1. In the MS, the first step is the identification of segments starting from the hits in each chamber. A muon track is reconstructed with a segment-seeded search (considering first the segments in the central layers of the detector as seeds, and then extending to the inner and outer ones). Except from the transition region between the barrel and the end-cap, tracks are required to have at least two matching segments; the hits associated to each track are fitted with a global χ^2, and the track candidate is accepted or rejected based on the χ^2 value.

Reconstructed muons can belong to four different types, depending on the sub-detectors that contribute to their reconstruction:

Combined   Combined muons are built from a global fit that uses hits from tracks reconstructed independently in the ID and MS.

Segment-tagged   Muons of lower p_T or that cross regions of lower acceptance of the MS can result in a segment in only one MS chamber. If a track from the ID is associated with this segment, this track is classified as a segment-tagged muon.

Calorimeter-tagged   A track in the ID is identified as a calorimeter-tagged muon if it is matched to an energy deposit in the calorimeter compatible with a minimum-ionizing particle. This muon category recovers identification efficiency for the muons that fall out of the MS acceptance.

Extrapolated   Extrapolated muons are reconstructed from tracks in the MS compatible with originating from the IP, to recover muons in high-η regions, outside of the ID acceptance.

When multiple types of muon are reconstructed for the same physical object, the redundant ones are removed with an overlap removal (OR) procedure that, when muons share the same ID tracks, gives priority to combined muons, and then to segment-tagged muons over combined muons. The OR with extrapolated muons gives preference to the MS track with the best quality.

5.4.2 Muon Identification

The reconstructed muons have to fulfill identification criteria that help reject the background constituted mostly by decays of pions and kaons. The variables used in the identification are the significance of the difference of the charge-over-momentum ratios measured by the ID and by the MS (q/p significance), the difference of the p_T measured in the ID ad in the MS divided by the p_T of the combined track (ρ'), the normalized χ^2 of the combined fit, and the number of hits in the different detector layers. Based on these quantities, four muon identification criteria are defined: Loose, Medium, Tight and High-p_T. Loose, Medium and Tight are inclusive categories with increasingly tighter requirements, while the High-p_T selection starts from the Medium selection and applies extra requirements that improve the momentum resolution for muons with p_T above 100 GeV.

The Medium identification criteria is the default in ATLAS, and is the one that minimizes the systematic uncertainties associated with the muon reconstruction and calibration. In the pseudorapidity region with $|\eta| < 2.5$, Medium muons are required to be combined muons with ≥ 3 hits in at least two MDT layers (except for muons

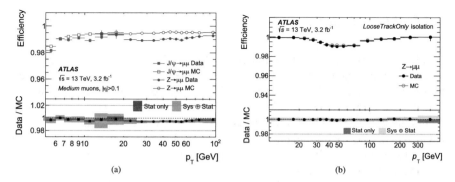

Fig. 5.9 **a** Reconstruction efficiency for the Medium muon selection as a function of the muon p_T, in the region with $0.1 < |\eta| < 2.5$, as derived from $Z \to \mu\mu$ and $J/\Psi \to \mu\mu$ events. **b** Efficiency of the muon isolation criteria for the *LooseTrackOnly* OP as derived from $Z \to \mu\mu$ events. In both figures, the top panel shows the efficiency for data and MC, while the bottom panes shows the ratio of data to MC with the corresponding uncertainty. Figures from Ref. [29]

with $|\eta| < 0.1$, in which case the requirement becomes on one MDT layer, but with a hole veto), while extrapolated muons are used when $2.5 < |\eta| < 2.7$. To suppress muons from hadron decays, the q/p significance is required to be less than seven.

5.4.3 Muon Efficiency Measurement

In the region with $|\eta| < 2.5$, where the information from both the ID and the MS is available, the reconstruction efficiency is measured with a tag-and-probe method, performed on $J/\Psi \to \mu\mu$ and $Z \to \mu\mu$ events for low-p_T and high-p_T muons respectively. After a selection on the event topology to reduce the background fraction, one of the two muons of the decay is required to be identified as a Medium muon (denoted as tag muon). The second leg of the decay (denoted as probe muon) has to be reconstructed by a system independent of the one to be calibrated. For example, calorimeter muons can be used to measure the efficiency of muon identification in the MS, while the ID efficiency can be measured with respect to muons identified in the MS. The reconstruction efficiency of the Medium OP as derived from $Z \to \mu\mu$ and $J/\Psi \to \mu\mu$ events is shown in Fig. 5.9a. The efficiency for high-η muons, for which the ID information is not available, is measured following the strategy detailed in Ref. [30].

5.4.4   Muon Isolation

While muons originating from semileptonic decays of hadrons are very close to the axis of a jet, prompt muons are typically well separated from other physics objects in the event (isolated). The isolation requirements help therefore suppressing further the background from semileptonically decaying hadrons. Different isolation requirements are available and calibrated in ATLAS, defined to be optimal for different analyses. The OP used in the analyses described in this thesis is the one labeled *LooseTrackOnly*, which applies a selection on the ratio of the variable $p_T^{varcone30}$ over the muon p_T (p_T^μ), where $p_T^{varcone30}$ is defined as the scalar sum of the momenta of the tracks with $p_T > 1$ GeV in the cone with $\Delta R < \min\left(10\,\text{GeV}/p_T^\mu, 0.3\right)$. The p_T-dependent size of the isolation cone helps in recovering efficiency for muons deriving from the decay of boosted particles. The efficiency of the isolation OPs is calibrated on $Z \rightarrow \mu\mu$ with the tag-and-probe method. The *LooseTrackOnly* OP has an efficiency of 99%, almost constant in η and p_T, as shown in Fig. 5.9b.

5.4.5   Muon Momentum Calibration

A set of corrections applied to the MC simulation, such that after the correction the simulation describes the muon momentum and momentum resolution in data with a precision of the order of few per-mill and few percent respectively. The corrections are derived in $J/\Psi \rightarrow \mu\mu$ and $Z \rightarrow \mu\mu$ events, by performing a binned maximum-likelihood fit of the dimuon invariant mass distribution.

5.5   Electrons

When electrons traverse the ATLAS detector, they leave a track in the ID and then the energy of their EM shower is absorbed in the ECal.

5.5.1   Electron Reconstruction

The electron reconstruction [31–33] starts with the identification of clusters in the ECal. The ECal can be divided in a grid of towers of size 0.025×0.025 in η and ϕ, and the tower energy is the sum of the energy of all the cells belonging to the tower. While, in the case of hadronic jets, the calorimeter clusters are created with a topological algorithm, in the case of electrons the calorimeter clusters are based on a sliding-window algorithm [6] with a size of 3×5 in units of 0.025×0.025 in the (η, ϕ) space, that searches for seed towers; these are the centers around which clusters are built.

The identified clusters are the starting point to reconstruct electrons, photons and converted photons. The key feature that allows to separate electrons from converted and unconverted photons is that, in the case of electrons, the calorimeter clusters are associated with a track from the ID; instead, in the case of converted photons the cluster is associated with a conversion vertex, while there is no track associated to an unconverted photon. Tracks from the ID are extrapolated to the second layer of the ECal, and a track is considered loosely matched to a seed cluster if the η difference between the track and the barycentre of the cluster is lower than 0.05 and the ϕ difference either lower than 0.2 (0.1 in the case of tracks deriving from hits only in the TRT) in the bending direction, or lower than 0.05 in the opposite direction. The tracks loosely matched with these criteria are then re-fitted with a Gaussian Sum Filter algorithm [34], that takes into account non-linear bremsstrahlung effects, and the re-fitted tracks are matched with the clusters with the same criteria as the loose matching, except from the ϕ difference in the direction of the bending, which is tightened to 0.1. If multiple tracks are associated to a cluster, only one is chosen as primary track based on the cluster-track distance. After the cluster-track matching, the cluster is re-built using groups of 3×7 (5×5) towers in the barrel (endcaps).

Once the electron is reconstructed, its energy is obtained from the energy of calorimeter cluster calibrated to the original electron energy with multivariate techniques [32], as will be discussed in Sect. 5.5.5, while the η and ϕ coordinates derive from the primary associated track.

Selections on the parameters of the primary track are applied to ensure that the electron is compatible with the PV interaction. In particular, in Run 2 analyses these selections are: $d_0/\sigma_{d_0} < 5$ and $z_0 \sin\theta < 0.5$ mm.

5.5.2 Electron Identification

After the electron candidates are reconstructed, a likelihood-based discriminant is used to reject the background, constituted mostly by hadronic jets and converted photons. This discriminant is built using as signal and background samples $Z \rightarrow ee$ and dijet events respectively for the high-E_T region, and $J/\Psi \rightarrow ee$ and minimum bias events respectively for the low-E_T region. Several variables are used to discriminate between signal and background, based on ratios of energy released in different layers of the calorimeter, shape of the EM shower, quality of the track and of the track-cluster matching; the full list of variables is reported in Table 5.2. The variables counting the number of hits in the different layers, as well as E/p, w_{stot} and $\Delta\phi_2$ are used to apply simple selections, while for the other discriminating variables PDFs are built based on the signal and background samples.

The product of these PDFs constitutes the signal and background likelihoods ($\mathcal{L}$s and $\mathcal{L}$s respectively), and the final discriminant is given by:

Table 5.2 Definitions of electron discriminating variables. Table from Ref. [33]

Type	Description	Name
Hadronic leakage	Ratio of E_T in the first layer of the hadronic calorimeter to E_T of the EM cluster (used over the range $|\eta| < 0 : 8$ or $|\eta| > 1 : 37$)	R_{had1}
	Ratio of E_T in the hadronic calorimeter to E_T of the EM cluster (used over the range $0.8 < |\eta| < 1.37$)	R_{had}
Back layer of EM calorimeter	Ration of the energy in the back layer to the total energy in the EM accordion calorimeter. This variable is only used below 100 Gev because it is known to be inefficient at high energies	f_3
Middle layer of EM calorimeter	Lateral shower width, $\sqrt{(\sum E_i \eta_i^2)/(\sum E_i) - ((\sum E_i \eta_i)/(\sum E_i))^2}$, where E_i is the energy and the η_i is the pseudorapidity of cell and i and the sum is calculated with in a window of 3×5 cells	w_{η^2}
	Ratio of the energy in 3×3 cells over the energy in 3×7 cells centered at the electron cluster position	R_ϕ
	Ratio of the energy in 3×7 cells over the energy in 7×7 cells centered at the electron cluster position	R_η
Strip layer of EM calorimeter	Shower width, $\sqrt{(\sum E_i (i - i_{max})^2)/(\sum E_i)}$, where i runs over all strips in a window of $\Delta\eta \times \Delta\phi \simeq 0.0625 \times 0.2$, corresponding typically to 20 strips in η, and i_{max} is the index of the highest-energy strip	w_{stot}
	Ratio of the energy dierence between the largest and second largest energy deposits in the cluster over the sum of these energies	E_{ratio}
	Ratio of the energy in the strip layer to the total energy in the EM accordion calorimeter	f_1
		(continued)

Table 5.2 (continued)

Type	Description	Name
Track conditions	Number of hits in the innermost pixel layer; discriminates against photon conversions	n_{Blayer}
	Number of hits in the pixel detector	n_{Pixel}
	Number of total hits in the pixel and SCT detectors	n_{Si}
	Transverse impact parameter with respect to the beam-line	d_0
	Significance of transverse impact parameter defined as the ratio of d_0 and its uncertainty	$d_0/\sigma d_0$
	Momentum lost by the track between the perigee and the last measurement point divided by the original momentum	$\Delta p/p$
TRT	Likelihood probability based on transition radiation in the TRT	e proba-bilityHT
Track–cluster matching	$\Delta\eta$ between the cluster position in the strip layer and the extrapolated track	$\Delta\eta_1$
	$\Delta\phi$ between the cluster position in the middle layer and the track extrapolated from the perigee	$\Delta\phi_2$
	Defined as $\Delta\phi_2$, but the track momentum is rescaled to the cluster energy before extrapolating the track from the perigee to the middle layer of the calorimeter	$\Delta\phi_{\text{res}}$
	Ratio of the cluster energy to the track momentum	E/p

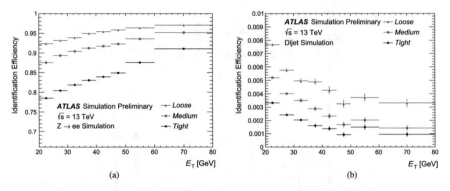

Fig. 5.10 **a** Electron identification efficiency in $Z \rightarrow ee$ events and **b** background mis-identification in dijet events for the three OPs Loose, Medium and Tight. Figures from Ref. [33]

$$d_{\mathcal{L}} = \frac{\mathcal{L}s}{\mathcal{L}s + \mathcal{L}b} \quad .$$

Three identification OPs are defined, Loose, Medium and Tight, optimized in bins of $|\eta|$ and E_T; these OPs are inclusive and with an increasing level of signal purity. The signal and background efficiency in MC samples is shown in Fig. 5.10. It is possible to notice how, with increasing E_T, the signal efficiency increases and the background mis-identification decreases.

5.5.3 Electron Isolation

As already discussed in the case of muons in Sect. 5.4.4, prompt signal electrons are in general more isolated than electrons candidates originating from hadron decays, from light hadrons misidentified as electrons or from converted photons. The analyses described in this thesis use a track-based isolation criterion, based on the variable $p_T^{varcone0.2}$, defined as the scalar sum of the p_T of the tracks satisfying quality requirements in a cone with $\Delta R < \min (10\,\text{GeV}/E_T, 0.2)$, where E_T is the transverse energy of the electron candidate and the sum excludes the electron track. The operating point used is *LooseTrackOnly*, that applies a selection on the ratio $p_T^{varcone0.2}/E_T$ to have an efficiency of 99% on simulated $Z \rightarrow ee$ events, constant as a function of E_T.

5.5.4 Electron Efficiency Measurement

The measurement of the electron efficiency in data relies on the tag-and-probe method, applied to $Z \rightarrow ee$ and $J/\Psi \rightarrow ee$ events, for the high-E_T (> 15 GeV)

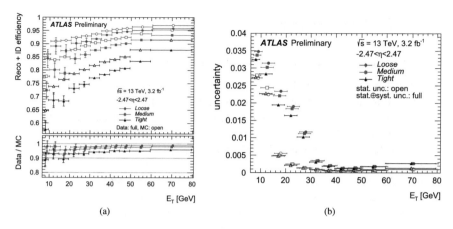

Fig. 5.11 **a** Combined electron reconstruction and identification efficiency in $Z \rightarrow ee$ simulated events and in the 2015 data and **b** absolute efficiency uncertainty, as a function of E_T and inclusive in η. Figures from Ref. [33]

and low-E_T (typically 7–20 GeV) regions respectively. One of the two electrons is identified with strict criteria and, after kinematic requirements, the second one is used to measure the efficiency. The electron efficiency is a product of the reconstruction, identification and isolation efficiency (and also trigger efficiency, if the events are selected with an electron trigger). The ratio between the efficiency expected from MC simulations and the one measured in data, in the form of SFs function of E_T and $|\eta|$, is used to correct the simulations and its uncertainty is applied as a systematic variation.

The identification efficiency is measured with four methods, always with respect to reconstructed electrons. Two methods, Z_{mass} and Z_{iso}, use $Z \rightarrow ee$ events. In the Z_{mass} analysis, the tag-probe invariant mass is required to be within 15 GeV of the Z boson mass, while in the Z_{iso} method the electron isolation is used to discriminate between signal and background. The other two methods [35], J/Ψ τ-cut and J/Ψ τ-fit, use the distribution of a variable related to the J/Ψ proper time (pseudo-proper time) to select ee events. The two Z-based methods and the two J/Ψ-based methods are combined, taking into account statistical and systematic correlations, to derive SFs in the high-E_T and low-E_T regions.

The reconstruction efficiency is measured as the ratio of reconstructed electrons to the number of EM clusters. It is measured with a method similar to the Z_{mass} method, but with the selection criteria for the probe relaxed to include all the EM clusters. Figure 5.11a shows the combined reconstruction and identification efficiency in $Z \rightarrow ee$ simulated events and in the 2015 data, as a function of E_T and inclusive in η, while Fig. 5.11b should the corresponding absolute uncertainty.

Also the isolation efficiency is measured with a tag-and-probe method derived from the Z_{mass} one, but with a lower E_T threshold for the probe electrons. For each isolation OP, the efficiency is derived with respect to each identification OP.

5.5.5   *Electron Energy Scale and Resolution*

The electron energy calibration follows three steps [32, 36]:

Detector response   Data-driven   corrections   are   derived   to   correct   for
   non-uniformity in the detector response, and are applied to data.

MC-based   The energy is corrected with a BDT that takes into account the energy
   deposited in front of the calorimeter (before reaching the first active layer of the
   calorimeter particles traverse 5–10 radiation lengths) and the changes in energy
   response depending on the impact point in the calorimeter. This calibration is
   derived from MC simulation and applied to both data and MC.

In-situ   After the application of the data-driven corrections for the detector non-
   uniformity and of the MC-based corrections, residual differences in electron
   energy scale and resolution between data and MC are measured with a template
   procedure on $Z \to ee$ events. The energy scale correction is applied to data, while
   an energy resolution smearing is applied to MC.

5.6   Missing Transverse Momentum

Particles that interact only weakly with the detector, such as neutrinos or BSM particles like neutralinos, are not reconstructed directly. Their presence is instead inferred
by measuring the total momentum imbalance in the event. The missing transverse
momentum vector ($\vec{E}_T^{\text{miss}}$) is defined as the negative vector sum of the p_T of all the
reconstructed calibrated objects in the event, plus a term that groups all the energy
that is not associated to any of the reconstructed objects (soft term) [37–39]. The
missing transverse momentum is given by:

$$\vec{E}_T^{\text{miss}} = -\sum \vec{p}_T^{\,e} - \sum \vec{p}_T^{\,\gamma} - \sum \vec{p}_T^{\,\tau} - \sum \vec{p}_T^{\,\text{jets}} - \sum \vec{p}_T^{\,\mu} - \sum \vec{p}_T^{\,\text{soft-terms}} .$$

The magnitude of $\vec{E}_T^{\text{miss}}$ is the missing transverse momentum (E_T^{miss}), and its
azimuthal angle is ϕ^{miss}.

   The soft term includes all the detector signals that are not associated to muons,
electrons, photons, taus or jets, and can receive contributions both from the hard
scattering and from pileup interactions. In ATLAS several algorithms are designed to
reconstruct and calibrate the E_T^{miss} soft term, and the analyses discussed in this thesis
use the one recommended for the 2015–2016 analyses, the track soft term (TST). In
this algorithm, the E_T^{miss} soft term is reconstructed purely from track information,
without any contribution from the calorimeter information; this results at the same
time in better pileup resistance but also in the loss of information about soft neutral
particles. An alternative version of the soft term is the calorimeter soft term (CST),
that instead uses energy deposits in the calorimeters not associated to hard physics
objects. Other algorithms are described in Ref. [37].

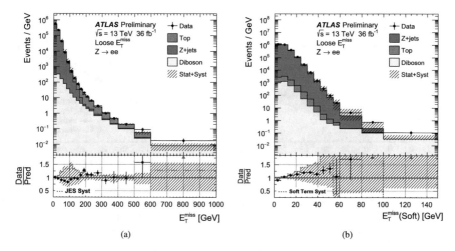

Fig. 5.12 Distribution of **a** E_T^{miss} using the Loose E_T^{miss} OP in data and simulation and **b** E_T^{miss} soft term in a $Z \to ee$ event selection. Figures from Ref. [39]

Three different E_T^{miss} OP are available, which differ in the selections on the jets that are used in the $\sum p_T^{jets}$ term in Eq. 5.4 [39]:

Loose Includes all jets with $p_T > 20$ GeV that pass the JVT selection when the jet has $p_T < 60$ GeV and $|\eta| < 2.4$. This is the OP used in the searches presented in this thesis.

Tight In addition to Loose criteria, the forward jets with $|\eta| > 2.4$ are required to have $p_T > 30$ GeV to be included in the E_T^{miss} computation.

Forward-JVT In addition to the Loose criteria, jets with $|\eta| > 2.5$, $p_T < 50$ GeV and failing the Loose fJVT criteria (described in Ref. [40]) are not included.

The performance of the E_T^{miss} reconstruction is evaluated in data and MC simulation, studying the mean, the width and the integral of the tail of the E_T^{miss} distribution in different topologies. In the 2015–2016 dataset, the E_T^{miss} performance has been evaluated using two different signatures. $Z \to \ell\ell$ events are studied both MC simulation and in data, since the leptonic decay of the Z boson are abundant and easy to trigger. These events do not contain any real E_T^{miss}, therefore all the reconstructed E_T^{miss} can be assigned to mismeasurement effects. The second signature used is vector boson fusion $h \to WW$ events, where both W bosons decay to a lepton and a neutrino; this topology is studied in MC simulation only.

The comparison of the E_T^{miss} distribution with the Loose OP and of the E_T^{miss} soft term in data and simulation is shown in Fig. 5.12a and b respectively, in a $Z \to ee$ event selection.

The E_T^{miss} resolution, defined as the root mean square (RMS) obtained from the combined distribution of the x and y components, respectively E_x^{miss} and E_y^{miss}, is shown in Fig. 5.13 for the Loose OP. The simulation agrees well with data within the uncertainties. In the definition of the resolution, the RMS is preferred over the

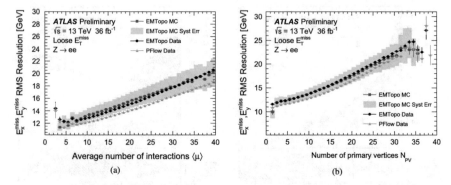

Fig. 5.13 The RMS obtained from the combined distributions of E_T^{miss} using the Loose E_T^{miss} OP for data with EMTopo jets (circular marker) and PFlow jets (triangular marker) and MC simulation with EMTopo jets (square marker) in a $Z \rightarrow ee$ event selection are shown versus **a** $<\mu>$ and **b** number of primary vertices. Figures from Ref. [39]

width of a Gaussian fit to the distribution to preserve the information on the tail. The systematic uncertainties in the energy scale and resolution of all the physics objects are propagated to the E_T^{miss} computation. The only systematic uncertainties affecting only E_T^{miss} are the ones related to the soft term, which are measured $Z \rightarrow \ell\ell$ events by comparing the simulation with data.

References

1. Cornelissen T, Elsing M, Fleischmann S, Liebig W, Moyse E, Salzburger A (2007) Concepts, design and implementation of the ATLAS new tracking (NEWT), Tech. Rep. ATL-SOFT-PUB-2007-007. ATL-COM-SOFT-2007-002, CERN, Geneva, Mar. https://cds.cern.ch/record/1020106
2. Kalman RE (1960) A new approach to linear filtering and prediction problems. Trans ASME–J Basic Eng 35
3. Fruhwirth R (1987) Application of Kalman filtering to track and vertex fitting. Nucl Instrum Meth A 262:444
4. ATLAS Collaboration (2015) Early inner detector tracking performance in the 2015 data at $\sqrt{s} = 13$ TeV. https://cds.cern.ch/record/2110140
5. Fruhwirth R, Waltenberger W, Vanlaer P (2007) Adaptive vertex fitting. J Phys G 34:N343
6. Lampl W et al (2008) Calorimeter clustering algorithms: description and performance. ATL-LARG-PUB-2008-002. https://cds.cern.ch/record/1099735
7. ATLAS Collaboration (2017) Topological cell clustering in the ATLAS calorimeters and its performance in LHC Run 1. Eur Phys J C 77:490. arXiv:1603.02934 [hep-ex]
8. Cacciari M, Salam GP, Soyez G (2008) The Anti-k(t) jet clustering algorithm. JHEP 0804:063. arXiv:0802.1189 [hep-ph]
9. ATLAS Collaboration (2017) Jet energy scale measurements and their systematic uncertainties in proton-proton collisions at $\sqrt{s} = 13$TeV with the ATLAS detector. Phys Rev D 96:072002
10. Cacciari M, Salam GP (2008) Pileup subtraction using jet areas. Phys Lett B 659:119. arXiv:0707.1378 [hep-ph]

11. ATLAS Collaboration (2013) Pile-up subtraction and suppression for jets in ATLAS. ATLAS-CONF-2013-083. https://cds.cern.ch/record/1570994
12. ATLAS Collaboration (2015) Jet energy measurement and its systematic uncertainty in proton–proton collisions at $\sqrt{s} = 7$ TeV with the ATLAS detector. Eur Phys J C 75:17. https://doi.org/10.1140/epjc/s10052-014-3190-y
13. ATLAS Collaboration (2015) Jet global sequential corrections with the ATLAS detector in proton–proton collisions at $\sqrt{s} = 8$ TeV. ATLAS-CONF-2015-002. https://cds.cern.ch/record/2001682
14. ATLAS Collaboration (2015) Data-driven determination of the energy scale and resolution of jets reconstructed in the ATLAS calorimeters using dijet and multijet events at $\sqrt{s} = 8$ TeV. ATLAS-CONF-2015-017. https://cds.cern.ch/record/2008678
15. ATLAS Collaboration (2015) Determination of the jet energy scale and resolution at ATLAS using Z/γ-jet events in data at $\sqrt{s} = 8$ TeV. ATLAS-CONF-2015-057. https://cds.cern.ch/record/2059846
16. ATLAS Collaboration (2016) Performance of pile-up mitigation techniques for jets in pp collisions at $\sqrt{s} = 8$ TeV using the ATLAS detector. Eur Phys J C 76:581. arXiv:1510.03823 [hep-ex]
17. ATLAS Collaboration (2014) Tagging and suppression of pileup jets with the ATLAS detector. ATLAS-CONF-2014-018. https://cds.cern.ch/record/1700870
18. Soyez G, Salam GP, Kim J, Dutta S, Cacciari M (2013) Pileup subtraction for jet shapes. Phys Rev Lett 110: 162001. arXiv:1211.2811 [hep-ph]
19. ATLAS Collaboration (2018) https://atlas.web.cern.ch/Atlas/GROUPS/PHYSICS/PLOTS/JETM-2016-011/. Accessed 2018-04-04
20. ATLAS Collaboration (2015) Selection of jets produced in 13 TeV proton–proton collisions with the ATLAS detector. ATLAS-CONF-2015-029. https://cds.cern.ch/record/2037702
21. Nachman B, Nef P, Schwartzman A, Swiatlowski M, Wanotayaroj C (2015) Jets from jets: re-clustering as a tool for large radius jet reconstruction and grooming at the LHC. JHEP 02:075. arXiv:1407.2922 [hep-ph]
22. ATLAS Collaboration (2011) Commissioning of the ATLAS high performance b-tagging algorithms in the 7 TeV collision data. ATLAS-CONF-2011-102. https://cds.cern.ch/record/1369219
23. Piacquadio G, Weiser C (2008) A new inclusive secondary vertex algorithm for b-jet tagging in ATLAS. J Phys Conf Ser 119:032032. http://stacks.iop.org/1742-6596/119/i=3/a=032032
24. ATLAS Collaboration (2015) Expected performance of the ATLAS b-tagging algorithms in Run-2. ATL-PHYS-PUB-2015-022. https://cds.cern.ch/record/2037697
25. ATLAS Collaboration (2016) Optimisation of the ATLAS b-tagging performance for the 2016 LHC Run. ATL-PHYS-PUB-2016-012. https://cds.cern.ch/record/2160731
26. ATLAS Collaboration (2016) Performance of b-jet identification in the ATLAS experiment. J Instrum 11:P04008. http://stacks.iop.org/1748-0221/11/i=04/a=P04008
27. ATLAS Collaboration (2018) Measurement of b-tagging efficiency of c-jets in $t\bar{t}$ events using a likelihood approach with the ATLAS detector. ATLAS-CONF-2018-001. https://cds.cern.ch/record/2306649
28. ATLAS Collaboration (2018) Calibration of light-flavour b-jet mistagging rates using ATLAS proton–proton collision data at $\sqrt{s} = 13$ TeV. ATLAS-CONF-2018-006. http://cdsweb.cern.ch/record/2314418
29. ATLAS Collaboration (2016) Muon reconstruction performance of the ATLAS detector in proton–proton collision data at $\sqrt{s} = 13$ TeV. Eur Phys J C 76:292. arXiv:1603.05598 [hep-ex]
30. ATLAS Collaboration (2014) Measurement of the muon reconstruction performance of the ATLAS detector using 2011 and 2012 LHC proton–proton collision data. Eur Phys J C 74:3130. arXiv:1407.3935 [hep-ex]
31. ATLAS Collaboration (2011) Expected electron performance in the ATLAS experiment. ATL-PHYS-PUB-2011-006. https://cds.cern.ch/record/1345327

32. ATLAS Collaboration (2014) Electron and photon energy calibration with the ATLAS detector using LHC Run 1 data. Eur Phys J C 74:3071. arXiv:1407.5063 [hep-ex]
33. ATLAS Collaboration (2016) Electron efficiency measurements with the ATLAS detector using the 2015 LHC proton–proton collision data. ATLAS-CONF-2016-024. https://cds.cern.ch/record/2157687
34. ATLAS Collaboration (2012) Improved electron reconstruction in ATLAS using the Gaussian Sum Filter-based model for bremsstrahlung. ATLAS-CONF-2012-047. https://cds.cern.ch/record/1449796
35. ATLAS Collaboration (2014) Electron efficiency measurements with the ATLAS detector using the 2012 LHC proton–proton collision data. ATLAS-CONF-2014-032. https://cds.cern.ch/record/1706245
36. ATLAS Collaboration (2016) Electron and photon energy calibration with the ATLAS detector using data collected in 2015 at $\sqrt{s} = 13$ TeV. ATL-PHYS-PUB-2016-015. https://cds.cern.ch/record/2203514
37. ATLAS Collaboration (2017) Performance of algorithms that reconstruct missing transverse momentum in $\sqrt{s} = 8$ TeV proton-proton collisions in the ATLAS detector. Eur Phys J C 77:241. arXiv:1609.09324 [hep-ex]
38. ATLAS Collaboration. Performance of missing transverse momentum reconstruction with the ATLAS detector using proton-proton collisions at $\sqrt{s} = 13$ TeV. arXiv:1802.08168 [hep-ex]. CERN-EP-2017-274
39. ATLAS Collaboration (2018) $E_{\mathrm{T}}^{\mathrm{miss}}$ performance in the ATLAS detector using 2015–2016 LHC pp collisions. ATLAS-CONF-2018-023. https://cds.cern.ch/record/2625233
40. ATLAS Collaboration (2017) Identification and rejection of pile-up jets at high pseudorapidity with the ATLAS detector. Eur Phys J C77(9):580. arXiv:1705.02211 [hep-ex]. [Erratum: Eur Phys J C77(10):712 (2017)]

Chapter 6
Statistical Methods

The result of a pp interaction is not deterministic, as the statements that can be obtained from quantum field theory are of probabilistic nature. Furthermore, uncertainties in our predictions related both to experimental effects and to the modeling of the physics processes need to be taken into account. Therefore, a proper statistical treatment is essential to extract quantitative statements from the observed data. This chapter discusses the main statistical methods that are used to obtain the results described in Chaps. 8 and 9. After a brief introduction to statistical inference in Sect. 6.1, the two main topics discussed are parameter estimation, in Sect. 6.2, used to determine the values of the model parameters that best describe data, and hypothesis testing, in Sect. 6.3, that checks the plausibility of models against the observed data. For illustration purposes, Sect. 6.4 describes a few simplified examples of applications of these statistical methods to physics analyses.

6.1 Statistical Inference

Statistical inference uses a data sample to make probabilistic statements on the population from which the sample is extracted. The generalization from the properties of the sample to the properties of the entire population comes with a certain degree of uncertainty, which can also be determined through statistical methods. There are two main approaches to statistical inference: frequentist and Bayesian.

Frequentist In the frequentist approach, probability is defined as the fraction of favorable outcomes of a repeatable experiment when the number of repetitions tends to infinity:

$$P(A) = \lim_{N_{tot} \to \infty} \frac{N_A}{N_{tot}} .$$

© Springer Nature Switzerland AG 2020
C. Rizzi, *Searches for Supersymmetric Particles in Final States*
with Multiple Top and Bottom Quarks with the Atlas Detector, Springer Theses,
https://doi.org/10.1007/978-3-030-52877-5_6

In this case there is no probability for a hypothesis or true values of the parameters: a theory is either true or false, and given a theory only the data has a certain probability or PDF.

Bayesian   In the Bayesian approach, probability is a more subjective notion, which incorporates the degree of belief in the form of priors. In this case it is meaningful to speak about the PDF not only for the data, but also for theories and parameters. Performing an experiment will modify the probability of a theory according to the Bayes formula:

$$P(\text{theory}|\text{data}) = \frac{P(\text{data}|\text{theory}) \times P(\text{theory})}{P(\text{data})},$$

where $P(\text{data}|\text{theory})$ is the probability of the data given the theory under examination, $P(\text{theory})$ is the theory prior, $P(\text{data})$ is a normalization constant that expresses the probability of observing these data whether the theory is true or false, and $P(\text{theory}|\text{data})$ is the final posterior probability of the theory.

In this chapter we discuss only about frequentist statistics, as only frequentist methods are applied to obtain the results presented in Chaps. 8 and 9.

As mentioned above, a hypothesis is a statement that we want to test against the observed data. A hypothesis can be simple, when the data PDF can be fully determined by stating if the hypothesis is true or false, or complex, when it depends on a number of parameters that in the following will be generally denoted with θ.

Given a certain hypothesis H, the probability of observing the data x is $P(x|H)$. This expression can be considered also as a function of the hypothesis, for fixed data, and in this case it is assigned the name of likelihood (L). In the case of a composite hypothesis, the value of the likelihood depends on the value of the parameters necessary to specify the hypothesis, and it takes the name of likelihood function: $L(\theta) = P(x|\theta)$. Note that while $P(x|\theta)$ as a function of x is a proper PDF and it normalizes to unity, this is not the case for the likelihood function, which depends on θ:

$$\int L(\theta) \, d\theta \neq 1 .$$

6.2   Parameter Estimation

Parameter estimation, also know as fitting, is the technique to derive from the data the best value of the model parameters, along with their uncertainties. This sections describes parameter estimation and in particular the method based on the maximum likelihood.

6.2.1 Estimators

Consider a set of N independent observations $\vec{x} = \{x_1, x_2, \ldots, x_N\}$, distributed accordingly to a PDF $f(x|\vec{\theta})$, where $\vec{\theta} = \{\theta_1, \theta_2, \ldots, \theta_M\}$ are the M true parameters that specify the underlying theory. A statistic is any function of the measured data $\vec{x}$, and an estimator $\hat{\vec{\theta}}$ is a statistic introduced to provide an estimate of $\vec{\theta}$. The characteristics used to evaluate the performance of an estimator are:

Consistency The asymptotic limit for large number of observations of the estimator is the true value of the parameter:

$$\lim_{N \to \infty} \hat{\vec{\theta}} = \vec{\theta} . \tag{6.1}$$

Bias For a given number N of observations, the bias is the difference between the expectation value of the estimator $(E[\hat{\vec{\theta}}])$ and the true value:

$$\vec{b} = E[\hat{\vec{\theta}}] - \vec{\theta} . \tag{6.2}$$

Once the bias of a specific estimator is known, it is possible to build an unbiased one by defining

$$\hat{\vec{\theta}}_{\text{unbiased}} = \hat{\vec{\theta}} - \vec{b} .$$

Efficiency The efficiency of an estimator is high when its variance, denoted as $V[\hat{\vec{\theta}}]$, is small. $V[\hat{\vec{\theta}}]$ has a lower bound, that is given by the Cramer-Rao inequality [1, 2].

Robustness An estimator is robust if it is not sensitive to small changes in the assumptions on the PDF $f(x|\vec{\theta})$, and is not excessively affected by outliers.

6.2.2 The Maximum Likelihood Estimator

The maximum likelihood estimate (MLE) of the parameters $\vec{\theta}$, first introduced by Fisher [3, 4], is obtained by choosing the set of parameters that yields the global maximum of the likelihood function. At an intuitive level this corresponds to the choice for which the observed data $\vec{x}$ are more probable.

The MLE is widely used because in the asymptotic limit of infinite number of events N this estimator is:

- Consistent, which means that it satisfies Eq. 6.1.
- Unbiased, so the bias, defined in Eq. 6.2, tends to zero as $N \to \infty$.
- Efficient, as it reaches the minimum bound for the variance.
- It follows a Gaussian distribution.

With a finite number of events, the estimator is biased and its bias is proportional to $1/N$. Also, the MLE is invariant under a functional transformation, which means that if $\hat{\theta}$ is the MLE for the parameter θ, $g(\hat{\theta})$ is the MLE for $g(\theta)$, and therefore it maintains all the properties described above. This is not necessarily true for other estimators.

Practically, most of the times instead of maximizing the likelihood function it is more convenient to minimize $-\log L$. This allows to perform a minimization rather than a maximization because of the minus sign, and gives the same result since the natural logarithm is a monotone transformation: as long as the likelihood is a monotonous function, its maxima are the same as the ones of the $\log L$. There are two types of maximum likelihood fits that can be performed:

Unbinned fits   Each event enters in the likelihood separately. This method is the optimal one from the statistics point of view, but it can be computationally expensive if the number of events is very large.

Binned fits   The events are grouped in bins of a histogram, and the quantity entering in the likelihood is the number of events in each bin.

In this chapter the focus will be on the second type of fit, as it is the one used in this thesis.

6.2.2.1   Variance of the Maximum Likelihood Estimator

The estimate of a parameter based on a data sample is always associated with a statistical uncertainty. If the same parameter estimate is performed on a different data sample, the values of the MLE are different; if this is repeated N times on independent data samples, for large N the best estimates will have a distribution that tends to a Gaussian, whose variance gives access to the statistical uncertainty on the MLE. Two ways to access this variance are:

- Once the parameter is estimated from the experimental results, the experiment can be simulated N times with MC simulations. For each simulation the best value of the parameter can be estimated with the maximum likelihood method, and from the distribution of these values it is possible to compute the variance.
- With the increase of the size of the data sample, the likelihood function tends to a Gaussian, and the $\log L$ tends to a parabola, so the variance of the distribution can be obtained as the point where $-\log L$ differs by 1/2 with respect to its minimum.

6.2.3   Binned Likelihood Fit with Systematic Uncertainties

If we consider the representation of the observed data given by the binned distribution of a certain variable, the expected number of events in each bin of the distribution is given by:

$$E[n_i] = b_i + \mu s_i,$$

where b_i is the expected number of background events, s_i the expected number of signal events, normalized to a given benchmark cross-section, and μ is a multiplicative factor of that cross-section referred to as signal strength. If the observed data follow a Poisson distribution, the likelihood is given by:

$$L(\mu) = \prod_{i=1}^{N} \frac{(\mu s_i + b_i)^{n_i}}{n_i!} e^{-(\mu s_i + b_i)}, \tag{6.3}$$

where the index i runs on all the bins of the distribution.

The prediction on the expected number of events in each bin is affected by systematic and statistical uncertainties, that are incorporated in the likelihood in the form of NPs. In the frequentist approach, each NP has a true unknown value, whose MLE is derived in auxiliary measurements. While in principle the full likelihood of the auxiliary measurements should be included in the likelihood of our own measurement, in most practical cases this is not feasible, and the auxiliary likelihood is modeled through constraining terms. Assuming that data used to derive the constraints on the NPs are statistically independent from the data in the analysis and not affected by the potential presence of the signal under study, the combined likelihood assumes the factorized form:

$$L(\mu, \vec{\theta}) = \prod_{i=1}^{N} \frac{(\mu s_i(\vec{\theta}) + b_i(\vec{\theta}))^{n_i}}{n_i!} e^{-(\mu s_i(\vec{\theta}) + b_i(\vec{\theta}))} \prod_{k=1}^{M} \rho(\theta_k),$$

where $\rho(\theta_k)$ is the constraint term for the parameter θ_k. Note that the treatment of an uncertainty through the addition of a constraint term, from the pure frequentist point of view is justified only for sources of uncertainty of experimental nature, where there is indeed an auxiliary measurement. The same approach is instead often also used for theoretical uncertainties, where the frequentist prescription would be to have a result that depends on these parameters. While in the Bayesian view the constraint term would easily be interpreted as a prior on the value of the theoretical parameter, in the frequentist approach this is equivalent to assigning also to the theoretical uncertainties a fictitious auxiliary measurement.

6.2.3.1 Functional Forms of the Constraint Terms

Depending on the underlying auxiliary measurement, different functional forms can be used for the constraint terms.

Gaussian constraint This is the default constraint type for systematic uncertainties, and its usage is justified by the central limit theorem. In general, it is safe to use this constraint as long as the MLE of the NP from the corresponding auxiliary measurement follows a Gaussian distribution. The functional form is:

$$\rho(\theta) = \frac{1}{\sqrt{2\pi}\sigma} \exp\left(-\frac{(\theta - \hat{\theta})^2}{2\sigma^2}\right) \ .$$

Technically, the function implemented in the likelihood is a truncated Gaussian, in order to avoid non-zero probability for non-physical values of the NP (e.g. values that correspond to a negative background prediction). This truncated Gaussian is normalized to still have unit area.

Poisson constraint   This constraint is often referred to as "Gamma" constraint because, when using it in a Bayesian approach, if combined with a uniform prior it gives rise to a Gamma posterior. It is used for NPs related to event count and to model the MC statistical uncertainty. If we assume that the background yield predicted from MC is n background events, derived from N MC simulated events with a multiplicative scale α, then the constraint term is:

$$\rho(n) = \frac{1}{\alpha} \frac{(n/\alpha)^N}{N!} \exp\left(-n/\alpha\right) \ .$$

Log-normal constraint   Log-normal is the distribution of a variable whose logarithm is distributed according to a Gaussian distribution. When random variables are multiplied, their product follows a log-normal distribution:

$$\rho(\theta) = \frac{1}{\sqrt{2\pi}\ln\sigma} \exp\left(-\frac{(ln\theta - \ln\hat{\theta})^2}{2(\ln\sigma)^2}\right) \frac{1}{\theta} \ .$$

One of the advantages of this constraint is that it never assumes negative values.

6.2.3.2   Effect of a Systematic Uncertainty on the Yield

Let's consider a single bin of a distribution and the effect that a systematic uncertainty, e.g. the JER, has on the expected background yield in this bin. The JER uncertainty is calibrated with an auxiliary measurement, whose likelihood we include by the simplification of a Gaussian constraint term, $G(\tilde{\theta}|\theta, \sigma_\theta)$, where $\tilde{\theta}$ is the nominal value of the calibration, θ the underlying real value and σ_θ the uncertainty. Stating that the uncertainty in JER is 10%, means that the width of the Gaussian is 10%.

To include this information in the likelihood, we need to evaluate the response to this uncertainty, i.e. the effect of a shift in JER on the background efficiency of our region. In practice this means that we need to evaluate how much a one standard deviation variation of JER (10% in our example) changes the number of expected events in the bin we are considering. For this example, we assume that a 10% shift in JER causes a 20% change in the background yields. If we consider only one bin of the likelihood in Eq. 6.3, and only one systematic (JER in this case), we write explicitly the effect of the systematic on the number of background events as:

$$L(\mu, \theta) = \frac{(\mu s + b \left(\frac{\theta}{\theta}\right) 2)^n}{n!} e^{-(\mu + b \left(\frac{\theta}{\theta}\right) 2)} \, G(\tilde{\theta}|\theta, \sigma_\theta) \ . \tag{6.4}$$

The change $b \rightarrow b\frac{\theta}{\theta}2$ encodes exactly the fact that a 10% change in JER leads to a 20% change in background yields. To simplify this expression, the auxiliary measurement can be normalized to a standard Gaussian: $G(\tilde{\theta}|\theta, \sigma_\theta) \rightarrow G(0|\theta, 1)$, where also θ has been normalized such that the values $\theta = \pm 1$ correspond to the nominal uncertainty. The likelihood then becomes:

$$L(\mu, \theta) = \frac{(\mu s + b(1 + 0.1\theta))^n}{n!} e^{-(\mu + b(1 + 0.2\theta))} \, G(0|\theta, 1) \ . \tag{6.5}$$

6.2.3.3 Interpolation

The dependence of the expected yield on a $\pm 1\sigma$ variation of each parameter in the likelihood is something that is measured in each specific analysis. This process is done only for three points (the nominal value and the $\pm 1\sigma$ variations), but it needs to be implemented in the likelihood as a continuous function. In the examples in Eqs. 6.4 and 6.5 a linear interpolation is assumed, but this is not always the best solution. Several interpolation strategies are possible; the ones described below and compared in Fig. 6.1 are those supported by HistFactory [5].

Piecewise Linear This is the simplest interpolation technique. The value of the predicted background yields is given by:

$$b(\theta) = \begin{cases} b + \theta(b^+ - b) & \theta \geq 0 \\ b + \theta(b - b^-) & \theta < 0 \end{cases} \tag{6.6}$$

where b is the nominal background prediction and $b^\pm$ correspond to the background yields for the $\pm 1\sigma$ variation of the NP. This interpolation technique has the clear advantage of simplicity, but it has the inconvenience of presenting a discontinuity at $\theta = 0$ when b^+ and b^- are not symmetric with respect to b as illustrated in Fig. 6.1b and d. For large uncertainties, it can also lead to negative $b(\theta)$, as shown in Fig. 6.1c.

Piecewise exponential The piecewise exponential interpolation is given by:

$$b(\theta) = \begin{cases} b \left(\frac{b^+}{b}\right)^\theta & \theta \geq 0 \\ b \left(\frac{b^-}{b}\right)^{-\theta} & \theta < 0 \end{cases} \ ,$$

In this case, $b(\theta)$ is bound to be positive. The discontinuity at $\theta = 0$ is still present, and in this case it appears also when b^+ and b^- are symmetric with respect to b (see Fig. 6.1c). Note that, once this is inserted into the likelihood, a Gaussian constraint with exponential interpolation is equivalent to a log-normal constraint with a linear interpolation. This is the default interpolation strategy used for the systematic uncertainties of the searches discussed in Chaps. 8 and 9.

Quadratic interpolation and linear extrapolation   In this case the interpolation between the $\pm 1\sigma$ response is given by the parabola passing through b^-, b^+ and b, while outside this range the extrapolation is given by the line tangent to the parabola in b^+ and b^-. This strategy avoids the discontinuity for $\theta = 0$, but it can introduce problems with the sign of the variation, for example sign inversion if $b - b^+$ and $b - b^-$ have the same sign, as shown in Fig. 6.1b.

Polynomial interpolation and exponential extrapolation   The   extrapolation outside the range of the $\pm 1\sigma$ response is given by the formula in Eq. 6.6, while the interpolation is obtained with a polynomial of sixth degree, bound to pass through b^-, b^+ and b and to match in value, first and second derivative the exponential extrapolation at the $\pm 1\sigma$ boundaries. This strategy avoids both the discontinuity at $\theta = 0$ and the possibility of negative $b(\theta)$.

6.2.3.4   Correlation and Profiling

The post-fit values of the NPs can be compared to the pre-fit values. A central value close to 0 and an uncertainty close to 1 indicates that the fit does not have enough statistical power to profile the uncertainties. If the central value is different from 0, it means that the best value is different from the nominal one; the modified MC prediction will have a better agreement with data than the original prediction. If the post-fit uncertainty on one NP is smaller than 1, it means that the original assigned uncertainty was too large and the fit was able to constrain the uncertainty on that NP.

When multiple NPs have a similar effect on the background prediction, the total variation obtained as a sum of their effects can be larger than what allowed by the statistical precision of the data. The individual effects cannot be disentangled and constrained individually, but a correlation between them produces a total variation that is compatible with what is observed in data.

6.2.4   Profiled Likelihood Ratio

Once we divide the parameters in the likelihood into parameter of interests (POIs), which are the ones we are interested in (typically the signal strength), and NPs, the likelihood itself can be maximized globally with respect to both the POI and the NPs $(L(\hat{\mu}, \hat{\theta}))$, but we can also find the conditional maximum for a certain value of the POI $(L(\mu, \hat{\hat{\theta}}))$. The ratio of these two quantities is the profiled likelihood ratio (PLR):

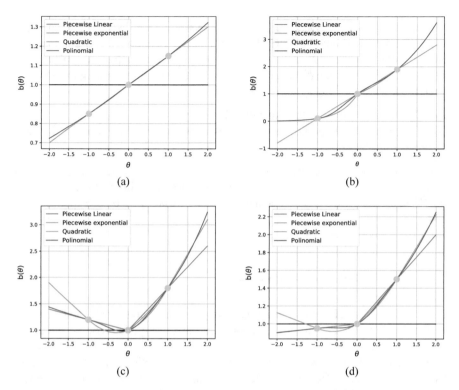

Fig. 6.1 Comparison of the four interpolation strategies supported by HistFactory [5]. θ represents the variation of the NP in units of standard deviations, and b(θ) the response relative to the nominal value; the three pink markers are the points where the response is measured (nominal and $\pm$ one standard deviation), while the colored lines show the interpolating function. **a** $\eta(-1) = 0.85$, $\eta(+1) = 1.15$. **b** $\eta(-1) = 0.1$, $\eta(+1) = 1.9$. **c** $\eta(-1) = 1.2$, $\eta(+1) = 1.8$. **d** $\eta(-1) = 0.95$, $\eta(+1) = 1.5$

$$\lambda(\mu) = \frac{L(\mu, \hat{\hat{\theta}})}{L(\hat{\mu}, \hat{\theta})} \ . \tag{6.7}$$

As it was discussed in Sect. 6.2.2, in order to determine the best-fit value of the POI it is more convenient to minimize the negative logarithm of the PLR, $-\log\lambda(\mu)$, rather than maximizing the PLR itself.

6.3 Hypothesis Testing

Hypothesis testing is the statistical procedure that allows to confirm or reject a specific model. In the case of high-energy physics, a typical example is the identification of the type of a particle based on its energy deposits in the detector. Another example, that will be be used as main test case in this section, is the discovery or exclusion of a BSM theory. When the alternative between two hypotheses is presented, they are typically referred to as null hypothesis and alternative hypothesis. If what we want to do is to proof the discovery of a new BSM signal by excluding the SM only hypothesis, the following symbols are used:

H_0 The null hypothesis, corresponds to SM only.
H_1 The alternative hypothesis, corresponds to the SM with the addition of the BSM process under test.

The test hypothesis can be generalized by including the signal strength μ, that acts as a cross-section modifier. The case $\mu = 1$ corresponds to the SM prediction plus the BSM process with its theoretical cross-section, while $\mu = 0$ corresponds to the SM.

6.3.1 Test Statistic and p-value

A test statistic t is a single real-value quantity, function of all the collected data. The PDF of the test statistic is different if we assume the null hypothesis or the alternative hypothesis to be true.

To quantify the compatibility of the observed data with a specific model, a p-value can be extracted from the distribution of the test statistic according to a certain hypothesis. For each hypothesis under test and given the observed value of the test statistic, t_{obs}, the p-value corresponds to the probability of having another observation more extreme than the current one:

$$p_\mu = \int_{t_{\text{obs}}}^{\infty} f(t|\mu)\, dt,$$

The p-value is therefore a frequentist statement on the conditional probability of having higher values for t if the measurement is repeated. When the p-value for the test hypothesis is lower than a predefined threshold, the test hypothesis is excluded. In high energy physics the convention is to set this threshold at 0.05. This corresponds to exclusion at 95% confidence level (CL). Note that excluding a test hypothesis does not mean confirming that the null hypothesis is correct, unless the union of the two covers all the possible phase space.

The threshold to exclude the null hypothesis (SM only) and declare a discovery has to be tighter than the one needed to exclude the test hypotheses (BSM). The 0.05 threshold would lead to 5% of the BSM searches to declare a discovery even in the absence of any real BSM signal. Instead, the convention is to declare evidence for New Physics when the p-value for the null hypothesis is lower than 1.3×10^{-3}, and discovery when it is lower than 2.9×10^{-7}.

Table 6.1 Conventional p-value and significance thresholds to exclude a test hypothesis, declare evidence or discovery of New Physics

	p-value	Z
Exclude test hypothesis	0.05	1.64
Evidence of New Physics	1.3×10^{-3}	3
Discovery of New Physics	2.9×10^{-7}	5

The significance Z of the p-value can be evaluated by transforming it in the equivalent number of standard deviations of a standard Gaussian needed to have an upper-tail integral equal to the p-value:

$$Z = \Phi^{-1}(1 - p),$$

where Φ is the cumulative of the standard Gaussian. Table 6.1 summarizes the conventional values for the exclusion of a signal hypothesis, the declaration of evidence or discovery of New Physics in terms of p-value and significance.

6.3.2 Test Statistic Using the PRL

While a test statistic can be any real-valued function of the data, the Neyman-Pearson lemma [6] ensures that the ones based on a likelihood ratio are the most statistically powerful. This means that, for a given signal efficiency, they provide the decision criterion that minimizes the misidentification probability. The PLR defined in Eq. 6.7 is the most used at the LHC, and it gives origin to this test statistic:

$$t_\mu = -2 \log \lambda(\mu),$$

and the corresponding p-value:

$$p_\mu = \int_{t_{\mu,\text{obs}}}^{\infty} f(t_\mu|\mu)dt_\mu \,.$$

Different variations of this test statistic are used for discovery and exclusion.

6.3.2.1 Test Statistic for Discovery

The discovery test statistic q_0 used to quantify the level of disagreement of data with the background-only hypothesis in case of an excess is defined as:

$$q_0 = \begin{cases} -2\ln\lambda(0) & \hat{\mu} \geq 0 , \\ 0 & \hat{\mu} < 0 . \end{cases} \tag{6.8}$$

The reason to assign the value 0 to the test statistic when $\hat{\mu} < 0$ is to avoid excluding the background-only hypothesis in case of a deficit. In fact, $\hat{\mu} < 0$ can indeed be symptomatic of a non correct background-only hypothesis (e.g. due to a systematic error), but it does not indicate the presence of a signal, which is what we want to highlight with this test statistic. Note that, since PLR assumes values between 0 and 1, q_0 is positive definite. The associated discovery p-value p_0 is:

$$p_0 = \int_{q_{0,obs}}^{\infty} f(q_0|0)\, dq_0 . \tag{6.9}$$

6.3.2.2 Test Statistic for Exclusion

When investigating the exclusion of the test hypothesis, we want a test statistic that does not penalize an excess. The exclusion test statistic q_μ is therefore defined as:

$$q_\mu = \begin{cases} -2\ln\lambda(\mu) & \hat{\mu} \leq \mu , \\ 0 & \hat{\mu} > \mu . \end{cases} \tag{6.10}$$

This test statistic, also known as one-sided PLR, is the default one used in the analyses described in Chaps. 8 and 9. The p-value q_μ is consequently defined as the integral above the observed value:

$$p_\mu = \int_{q_{\mu,obs}}^{\infty} f(q_\mu|\mu)\, dq_\mu .$$

Note that switching from the discovery test statistic to the exclusion test statistic is equivalent to inverting the role of the background-only and signal-plus-background hypothesis: in the case of the exclusion fit, the null hypothesis to exclude is the signal-plus-background one.

6.3.2.3 Uncapped Test Statistic

The strategy described above leads to a loss of information when the test statistic is set to 0. A solution is obtained by uncapping the test statistic and, instead of assigning a 0 to the situations we do not want to penalize, assigning them a negative value. This is achieved with the r_0 and r_μ test statistic. As an example, the definition for r_μ is:

$$r_\mu = \begin{cases} -2\ln\lambda(\mu) & \hat{\mu} \leq \mu , \\ +2\ln\lambda(\mu) & \hat{\mu} > \mu . \end{cases}$$

6.3.2.4 Allow only Positive Signals

The alternate PLR, $\tilde{\lambda}(\mu)$, is designed to take into account the fact that, in most cases, only positive μ have a physical meaning. In this case, when $\hat{\mu} < 0$, the best physical value is 0, and the alternate PLR is defined as:

$$\tilde{\lambda}(\mu) = \begin{cases} \frac{L(\mu, \hat{\hat{\theta}}(\mu))}{L(\hat{\mu}, \hat{\theta})} & \hat{\mu} \geq 0, \\ \frac{L(\mu, \hat{\hat{\theta}}(\mu))}{L(0, \hat{\hat{\theta}}(0))} & \hat{\mu} < 0 . \end{cases}$$

The corresponding test statistics are indicated with $\tilde{q}$ and are defined as in Eqs. 6.8 and 6.10 after substituting $\lambda(\mu) \rightarrow \tilde{\lambda}(\mu)$.

6.3.3 The CL$_S$ Method

If we consider a situation where H_0 and H_1 give similar distribution of the exclusion test statistic (e.g. because the signal cross-section is small) and we observe a downward fluctuation in data, we could end up excluding the test hypothesis, even if almost indistinguishable from the null hypothesis. The modified confidence level (CL$_s$) method [7] recovers from this situations by defining:

$$CL_s = \frac{q_\mu}{1 - p_b}, \qquad (6.11)$$

where, as shown in Fig. 6.2:

$$p_b = \int_0^{q_{\mu,\mathrm{obs}}} f(q_\mu | 0) \, dq_\mu .$$

With the CL$_s$ method, the test hypothesis is excluded at 95% CL if CL$_s$ < 0.05. In the situation described above, where the null and the test hypotheses give similar q_μ distribution, in case of a deficit both the numerator and the denominator in Eq. 6.11 will be small, and the test hypothesis will not be excluded. The CL$_s$ method is the default procedure used in the searches discussed in this document to decide on the exclusion of a signal model.

6.3.4 Distribution of the Test Statistic

To compute the p-value associated with the observed test statistic t, we need the PDF of t assuming that the signal hypothesis H_μ is true, $f(t|\mu)$. The distribution

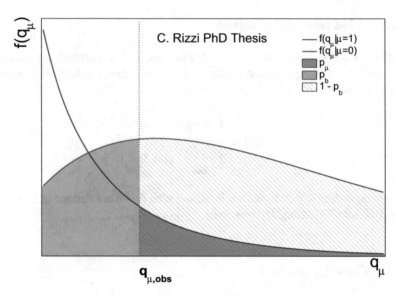

Fig. 6.2 Distribution of the test statistic q_μ in the case of background only (red line) and signal-plus-background (blue line) hypothesis. The filled blue are indicates the value of p_μ, while the filled red area indicated the value of p_b

of the test statistic can be obtained with pseudo-experiments or, in the case of large statistics, with the asymptotic approximation.

6.3.4.1 Pseudo-experiments

The distribution of the test statistic can be obtained by sampling the likelihood function with MC simulations (pseudo-experiments). This is obtained by repeating many times the following steps:

- The nominal value of all the NPs is varied by sampling randomly their constraint terms.
- The new value of the NPs is used to compute a new expected value for the yields.
- The observed value is substituted by a Poisson fluctuation of the expected yields.
- The test statistic for this "observed" value is computed.

This is done twice: a first time using as expected number of events the one predicted by the null hypothesis, and a second time the one predicted by the test hypothesis. The integral of the two resulting distributions is used to compute the p-values.

6.3.4.2 Asymptotic Approximation

For large number of events, the asymptotic approximation [8] can be used to determine the PDF of the test statistic without having to simulate a large number of pseudo-experiments. This technique is based on Wald's theorem [9], which states that $t_\mu = -2 \ln \lambda(\mu)$ is parabolic up to corrections that scale with the inverse of the square root of the sample size:

$$t_\mu = -2 \ln \lambda(\mu) = \frac{(\mu - \hat\mu)^2}{\sigma^2} + \mathcal{O}(1/\sqrt{N}) .$$

Ignoring the $\mathcal{O}(1/\sqrt{N})$ terms, $t_\mu = -2 \ln \lambda(\mu)$ is then distributed according to a noncentral chi-square distribution with one degree of freedom:

$$f(t_\mu; \Lambda) = \frac{1}{2\sqrt{t_\mu}} \frac{1}{\sqrt{2\pi}} \left[\exp\left(-\frac{1}{2} \left(\sqrt{t_\mu} + \sqrt{\Lambda} \right)^2 \right) + \exp\left(-\frac{1}{2} \left(\sqrt{t_\mu} - \sqrt{\Lambda} \right)^2 \right) \right] ,$$

(6.12)

where the noncentrality parameter Λ is:

$$\Lambda = \frac{(\mu - \mu')^2}{\sigma^2} .$$

The value of σ can be estimated through the Asimov dataset, defined as the dataset that, when used to estimate the likelihood parameters, leads to their true values. In practice the Asimov dataset is built by setting data equal to the nominal background prediction. Plugging this back into Eq. 6.12 allows to obtain a functional form of the distribution of the test statistic. As discussed in Ref. [8], even if the approximations used to derive the asymptotic formulae hold in the case of a large data sample, the agreement between this method and pseudo-experiments is at the level of $\approx 10\%$ even for an expected background of three events.

6.4 Simplified Examples

In this section we present a few simplified examples that illustrate the concepts described in the previous sections and how they are applied in physics analyses that search for BSM signals. In particular, Sect. 6.4.1 describes the guideline followed in optimizing signal regions (SRs), Sect. 6.4.2 discusses the advantages of using control regions (CRs), while Sects. 6.4.3 and 6.4.4 focus respectively on limit setting and on how to improve limits by combining several SRs. In all the examples in this section, as well as in the searches described in Chaps. 8 and 9, the profile likelihood fit is implemented in the HistFitter framework [10].

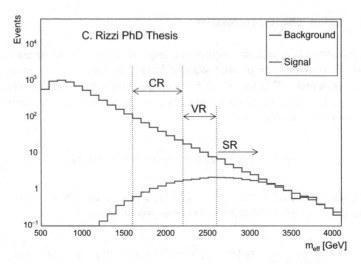

Fig. 6.3 Distribution of a discriminating variable (m_{eff}) for the signal and background in the example in the text. The arrows indicate the SR defined by maximizing the expected significance and the CR where the background is normalized

6.4.1 Region Definition

Let's consider a situation where we want to use only one discriminating variable to separate signal and background, e.g. m_{eff}, defined as the scalar sum of the momenta of all the objects in the event and missing transverse momentum, and they are distributed as in Fig. 6.3. We can see that the background, which in this case we consider as constituted by a single physical process, tends to have lower values of m_{eff} with respect to the signal; this means that the SR will be defined by a lower boundary on the value of m_{eff}. There are different criteria that can be used to choose the value of m_{eff} that defines the SR. The one used in this document is the maximization of the expected significance, computed through the function BINOMIALEXPZ in ROOSTATS [11], assuming a 30% uncertainty on the background yields. With this definition, the significance is the equivalent number of standard deviations of a Gaussian of a p-value defined as:

$$p(N_s, N_b, \sigma_b) = \mathcal{I}\left(\frac{1}{1+\frac{1}{N_b\sigma_b^2}}; N_s + N_b; \frac{1}{\sigma_b^2} + 1\right), \tag{6.13}$$

where N_s (N_b) is the expected number of signal (background) events, and σ_b the relative uncertainty on the background yields, and $\mathcal{I}$ the regularized beta function.

In the case of the signal and background models shown in Fig. 6.3, values of m_{eff} between 2 and 4 TeV have been tested in steps of 100 GeV, and the optimal value has been found to be $m_{\mathrm{eff}} > 3000$ TeV. This selection leads to a SR with 9.5 signal

Table 6.2 Pre-fit background yields, signal yields, signal-to-background ratio, significance and pseudo-data yields in the regions used in the examples in the text

	CR	VR	SR
Background	279.1	62.9	10.5
Signal	6.3	14.6	9.5
S/B [%]	2.2	23	91
Significance	0	0.48	1.58
Pseudo-data	231	52	15

events, 10.5 background events and an expected significance of 1.58 σ. The portion of the m_{eff} spectrum that is not occupied by the SR, can be used to define a CR and a validation region (VR). In this example we use the region with m_{eff} between 1.6 and 2.2 TeV as CR for the background. In order to evaluate the effect of the inclusion of a CR in the background estimate, we need to compare the number of expected and observed background events in the CR. In these examples, the data sample is substituted by a pseudo-data sample generated with a Poisson fluctuation of the bin-content of a histogram built as the sum of the background histogram with a scale factor of 0.87 and the signal histogram with a scale factor of 0.5. The summary of the pre-fit background yields, signal yields, signal-to-background ratio, significance and pseudo-data yields in all the regions is given in Table 6.2.

6.4.2 Background-Only Fit and Advantages of the Control Regions

CRs are used first of all in the so called "background-only fit", where the likelihood comprises only the CRs and any signal contamination is neglected. The goal of the background-only fit is to extract from the data in the CRs information on the modeling of the background and extrapolate it to VRs and SRs. A background estimate based on a CR has two main advantages with respect to the nominal MC prediction:

- First of all, the usage of a CR allows to eliminate mis-modellings in the normalization of the background sample in a phase space kinematically close to the SR. This is implemented through the inclusion in the likelihood of a normalization NP with a flat constraint. The improvement in the description of the data in the VR is shown in Fig. 6.4: the central value of the bottom panel, showing the ratio of the pseudo-data to the MC simulation, is closer to one after the fit.
- The inclusion of a CR in the fit plays a key role also in the evaluation of the systematic uncertainties. For a background normalized in a CR, the impact of a systematic uncertainty in a SR or VR does not depend on the full change in yields between the nominal and the systematic variation in that SR or VR, but it depends only on the change in the transfer factor (TF), defined as the ratio of the predicted

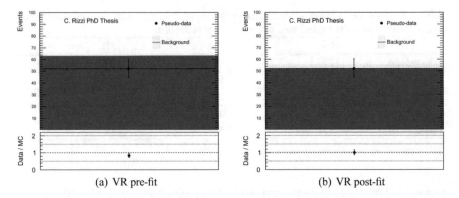

Fig. 6.4 Agreement between MC simulation and pseudo-data **a** before and **b** after the fit in the CR. No systematic uncertainties are included in the fit

yields in each SR/VR to the yields in the CR. Let's include in the previous example a systematic variation that impacts both the shape and the normalization of the m_{eff} distribution: we define a systematic uncertainty that increases the normalization of the sample by 20% and also changes the shape of the m_{eff} distribution by assigning to each event a weight corresponding to the m_{eff} value (expressed in GeV) divided by 20000; the uncertainty is then symmetrized.

This leads to the following yields in the CR and VR:

CR   Nominal: 279.1, up variation: 360.3, down variation: 197.8.
VR   Nominal: 62.9, up variation: 83.2, down variation: 42.5.

When a background is normalized in a CR, the size of the systematic uncertainty is determined by the relative change in the TF and not by the plain change in yields. This is because the normalization of the background in the CR absorbs the normalization uncertainty, and the effect of the uncertainty on the VR depends only on the change in shape of the distribution. In this example, while the pre-fit uncertainty in the VR is large, the variation of the transfer factor is significantly smaller:

$$\text{TF}_{\text{VR}}^{\text{nominal}} = \frac{62.9}{279.1} = 0.225,$$

$$\text{TF}_{\text{VR}}^{\text{up}} = \frac{83.2}{360.3} = 0.231,$$

$$\text{TF}_{\text{VR}}^{\text{down}} = \frac{42.5}{197.8} = 0.215.$$

This leads to an overall reduction of the impact of the systematic uncertainty, as shown in Fig. 6.5: after the fit in the CR, the shaded band indicating the uncertainty on the background estimate is smaller. This example was carried out on a VR, but the same is true for the uncertainty in the SR.

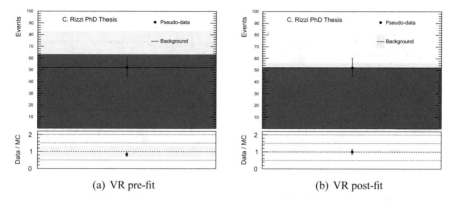

Fig. 6.5 Agreement between MC simulation and pseudo-data **a** before and **b** after the fit in the CR. The large pre-fit systematic uncertainty is reduced by the fit in the CR

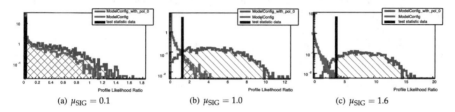

Fig. 6.6 Distribution of the test statistic for the background-only (blue) and signal-plus-background (red) hypothesis for **a** $\mu_{SIG} = 0.1$, **b** $\mu_{SIG} = 1.0$, **c** $\mu_{SIG} = 1.6$

6.4.3 Inclusion of the Signal Region in the Fit and Limits

Once the background model has been derived with the fit in the CR and validated in the VR, it is possible to set model-dependent limits by including also the SR in the likelihood ("exclusion fit").

First of all, the CL_s for the nominal signal cross-section is computed based on the distribution of the exclusion test statistic q_μ (described in Sect. 6.3.2). This can be done using the PDF for q_μ, obtained with pseudo experiments, as shown in Fig. 6.6b, or with the asymptotic approximation.

A second step consists in deriving an upper limit (UL) on the signal strength μ_{SIG}, which means finding by how much the nominal cross-section of the signal should be scaled to find a CL_s of exactly 0.05. This is obtained by performing multiple times the same hypothesis test just described, each time scaling the signal by a different factor. Also in this case, each point of this scan can be obtained either with the asymptotic approximation or with pseudo-experiments (some examples are shown in Fig. 6.7); note that in this case the "data" sample we are using is a pseudo-data sample built from a Poisson sampling of the sum of background scaled by 0.87 and signal scaled by 0.5, so the observed value of the test statistic is "signal-like". The interpolation of

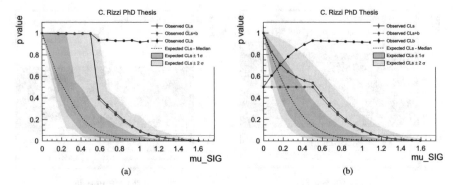

Fig. 6.7 Scan of the CL$_s$ for different values of the signal strength with **a** 5000 pseudo experiments and **b** the asymptotic approximation

Table 6.3 Expected and observed upper limit on μ_{SIG} for the example in the text, computed with the asymptotic approximation and with 5000 pseudo-experiments

	Observed	Expected	-1σ	$+1\sigma$	-2σ	$+2\sigma$
Asymptotic	1.15	0.69	0.48	1.01	0.35	1.43
Pseudo-experiments	1.14	0.71	0.51	0.99	0.40	1.35

these CL$_s$ values allows to determine the signal strength where CL$_s$ = 0.05; this is shown in Fig. 6.7a, where the p-values are determined based on pseudo-experiments, and in Fig. 6.7b, where instead the asymptotic approximation is used. The numerical UL values resulting from these CL$_s$ scans are reported in Table 6.3.

6.4.4 Improving the Sensitivity by Combining Regions

In this section we show how combining different regions can increase the sensitivity. For this we use the same signal sample as in the previous examples. The m_{eff} value that defines the SR has been determined by optimizing the expected significance. Since, in this case, the signal has a harder m_{eff} spectrum than the background, the phase-space region with m_{eff} below this threshold has a lower signal fraction. Nevertheless, it can still provide information useful to discriminate signal and background. If we consider for example the region with m_{eff} between 2.6 and 3.0 TeV (which is non-overlapping with the previous SR definition), here we expect 17.1 background events and 7.4 signal events; assuming a 30% uncertainty on the background estimate, this corresponds to an expected significance of 0.82 standard deviations, far below the significance of 1.58 standard deviations in the SR previously defined. Note that this region overlaps with the VR defined in Sect. 6.4.1, which in the case of this last example would need to be modified.

If we include this new region as well into the likelihood fit and perform an upper-limit scan (in this case, for simplicity, only with the asymptotic approximation), we have that the expected UL on μ_{SIG} is 0.62. This corresponds to an improvement of about 10% on the expected UL, obtained just by adding to the fit a region with limited sensitivity. This effect is increased when the regions that are considered simultaneously in the fit all offer a good sensitivity and have different signal fractions. This additional SR would play a bigger role for signal models giving a less hard m_{eff} spectrum, which are therefore less trivial to discriminate from the background with a single selection.

References

1. Cramér H (1946) Mathematical methods of statistics
2. Rao CR (1945) Information and the accuracy attainable in the estimation of statistical parameters. Springer New York, New York, NY. https://doi.org/10.1007/978-1-4612-0919-5_16
3. Fisher R, Pearson K (1911) On an absolute criterion for fitting frequency curves. Messenger of mathematics. https://books.google.fr/books?id=dXXzjgEACAAJ
4. Aldrich J (1997) R.A. Fisher and the making of maximum likelihood 1912–1922. Stat Sci 12:162
5. Cranmer K, Lewis G, Moneta L, Shibata A, Verkerke W (2012) HistFactory: a tool for creating statistical models for use with RooFit and RooStats. Tech. Rep. CERN-OPEN-2012-016, New York U., New York, Jan. https://cds.cern.ch/record/1456844
6. Neyman J, Pearson ES (1933) On the problem of the most efficient tests of statistical hypotheses. Philos Trans R Soc Lond Ser A, Containing Papers of a Mathematical or Physical Character 231:289. http://www.jstor.org/stable/91247
7. Junk T (1999) Confidence level computation for combining searches with small statistics. Nucl Instrum Meth A 434:435. arXiv:hep-ex/9902006 [hep-ex]
8. Cowan G, Cranmer K, Gross E, Vitells O (2011) Asymptotic formulae for likelihood-based tests of new physics. Eur Phys J C 71:1554. arXiv:1007.1727 [physics.data-an]. Erratum: Eur Phys J C 73:2501 (2013)
9. Wald A (1943) Tests of statistical hypotheses concerning several parameters when the number of observations is large. Trans Am Math Soc 54:426
10. Baak M, Besjes GJ, Côte D, Koutsman A, Lorenz J, Short D (2015) HistFitter software framework for statistical data analysis. Eur Phys J C 75:153. arXiv:1410.1280 [hep-ex]
11. Moneta L, Belasco K, Cranmer KS, Kreiss S, Lazzaro A, Piparo D, Schott G, Verkerke W, Wolf M (2010) The RooStats project. PoS ACAT2010:057. arXiv:1009.1003 [physics.data-an]

Chapter 7
Common Aspects to SUSY Searches with Multiple b-jets and E_T^{miss}

The main results presented in this thesis are the two analyses described in Chaps. 8 and 9, which target respectively the strong and electroweak production of super-symmetric particles. While these analyses target two different benchmark models, they have many commonalities, since both models feature a final state rich in b-jets and E_T^{miss}. This chapter highlights these common aspects: Sect. 7.1 illustrates the data sample used in the analyses. Section 7.2 describes the philosophy of simpli-fied models, used to design and interpret the searches. Sections 7.4 and 7.5 focus on the definition of the physics objects and the experimental systematic uncertainties respectively; Sect. 7.6 presents the main kinematic variables used in the analyses while Sect. 7.7 describes the background sources from SM processes, how they are modeled in the analyses and the uncertainties on this modeling.

7.1 Data Sample

The data used in this thesis are pp collisions produced by the LHC at a center-of-mass energy $\sqrt{s} = 13$ TeV during 2015 and 2016 and collected by the ATLAS detector, corresponding to an integrated luminosity of 36.1 fb^{-1} after the application of the GRL. The gluino search described in Chap. 8 has also been updated with the data collected in 2017, and in this case the total dataset has an integrated luminosity of 78.9 fb^{-1}.

Events are selected with E_T^{miss} triggers. The version of E_T^{miss} used in the trigger is calorimeter-based, which means that the muons are not included in the computation of E_T^{miss} and are therefore treated as invisible particles. As described in Sect. 3.3.7, trigger chains for physics analyses comprise first a L1 trigger selection, and then a selection from the HLT.

© Springer Nature Switzerland AG 2020
C. Rizzi, *Searches for Supersymmetric Particles in Final States with Multiple Top and Bottom Quarks with the Atlas Detector*, Springer Theses,
https://doi.org/10.1007/978-3-030-52877-5_7

L1   At L1 level, the computation of the E_T^{miss} is based on trigger towers of size
0.1 × 0.1 in the η, ϕ space. The towers are clustered in 4 × 4 group with a sliding-
window algorithm, and are used in the E_T^{miss} computation.

HLT   Many HLT E_T^{miss} algorithms are available in ATLAS and are described in
Ref. [1]. The one used in this thesis is the jet-based algorithm, where E_T^{miss} is
computed as the negative sum of the p_T of the trigger-level jets in the event.

For each run, the E_T^{miss} trigger with the lowest available unprescaled threshold
is used; this corresponds to an HLT threshold of 70 GeV, 90 GeV, 100 GeV and
110 GeV for the 2015, early 2016, mid 2016 and late 2016/2017 respectively. These
triggers are fully efficient after requiring that the offline reconstructed E_T^{miss} exceeds
200 GeV and that there are at least four jets in the event; these selections are applied
in all the regions of the searches in this thesis.

Beside the trigger and GRL requirements, events are required to have a reconstructed primary vertex with at least two associated tracks with $p_T > 0.4$ GeV.

7.2   Simplified Models

Even with the simplifying assumptions of the pMSSM, discussed in Sect. 2.3.6, a
SUSY model has to take into account a large number of free parameters, whose values
can impact the characteristics of the particle production and decay. Simplified models
[2] are very simple models of BSM Physics involving only a few particles and decay
modes, each one focusing on a specific signature, which can be used to optimize
and interpret analyses targeting BSM scenarios. In general simplified models can
be viewed as a limit of more complete models, where all particles except a few are
too heavy to be produced in the interactions. This leads to a drastic reduction of the
number of parameters: a simplified model can be described just by the production
cross-section and mass of the few particles considered. The BRs of the new particles
can be a parameter of the simplified model as well, but it is in general easier to consider
each decay chain as a separate simplified model with 100% BR. When simplified
models are used to discover or exclude a certain topology, it is important to connect
the results obtained to more general models, which can be done e.g. by relaxing the
restrictions on the BR of the BSM particles or allowing additional particles to take
part in the interaction.

7.3   Analysis Strategy

The analysis strategy is based on the definition of so-called SRs, signal-enriched
regions defined by selections on relevant kinematic variables with the goal of maximizing the sensitivity to specific benchmark models. After the SRs are defined, the
key aspect of the analysis is an accurate estimate of the number of events expected

Fig. 7.1 Schematic view of
an example relation between
CR, SR and VRs

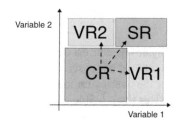

from SM processes (background events) in these regions, and of the associated uncertainty. With the background estimate at hand, it is possible to look at the observed yields in data and compare it with the expectations; this comparison, performed with the statistical methods discussed in Chap. 6, allows to quantify the significance of an excess or to place limits on BSM signal models.

The estimate of the expected background events can be obtained with different techniques. In particular, in these thesis three types of techniques are used:

- A first option is to take the background estimate directly from MC simulation.
- The MC estimate can be improved by normalizing each process in a dedicated CR, which is a region non overlapping with the SR, but kinematically close to it, enriched in the background process that we want to normalize and with a very low expected signal fraction. Since CR and SR must be non-overlapping, some of the selections of the SR are inverted to design the CR; the extrapolation of the background normalization factor from the CR to the SR is tested in dedicated VRs. Figure 7.1 shows a schematic view of the relation between CRs, VRs and VRs. A simplified example of how the usage of CRs can lead to an improved background prediction with reduced systematic uncertainties is discussed in Sect. 6.4.2.
- In some cases, a background estimate that relies only on data and not on MC simulation is preferred. In these cases we speak of a data-driven background estimate.

In each of the analyses described in this thesis, two different analysis strategies are carried out in parallel:

Cut-and-count   Several SRs are designed, each optimized to maximize the discovery significance to a specific region of the parameter space, represented by one benchmark model. Cut-and-count SRs are useful also to provide simple and powerful model-independent ULs, that are easy to reinterpret for signal models different from the ones considered in the analysis.

Multi-bin   In this case, the SRs are non-overlapping. This requirement conflicts with the simple choice of the best selection to maximize the significance, which means that the individual discovery power of each SR is smaller than in the case of cut-and-count SRs. On the other hand, having non-overlapping SRs allows to statistically combine them in the likelihood fit, leading to a stronger model-dependent expected UL. The increase in expected exclusion obtained with the combination of several regions is demonstrated in Sect. 6.4.4.

7.4 Object Definition

Physics objects used in the analyses are defined with two sets of inclusive selections. A first set of looser selections defines the baseline objects; these are used as input for the overlap removal algorithm which, in the steps defined in the next paragraph, solves the possible double-counting of physics objects (e.g. electrons reconstructed also as jets). Objects that survive the overlap removal procedure are then subject to tighter selections to define the signal objects. The specific selections applied to jets, electrons and muons are:

Jets   Baseline jets are reconstructed with the anti-k_T algorithm with radius parameter $R = 0.4$ and use the calibration procedure discussed in Sect. 5.2.3. The kinematic selection on baseline candidate jets is $p_T > 20$ GeV and $|\eta| < 2.8$; In order to suppress fake jets originating from pileup interactions, jets with $p_T < 60$ GeV are required to have JVT > 0.59 and satisfy the cleaning criteria discussed in Sect. 5.2.5 with the *BadLoose* OP. After overlap removal, signal jets are required to have $p_T > 30(20)$ GeV for the analysis discussed in Chaps. 8 and 9; the difference on the p_T threshold in the two analyses is related to differences in the signal event kinematics, and will be discussed more in detail in the specific chapters.

Jets are considered to be b-tagged based on the output of the MV2c10 algorithm (see Sect. 5.3); in both analyses the OP used corresponds to an average efficiency of 77% for b-jets from simulated $t\bar{t}$ events, and to a rejection factor of 6, 22 and 134 against jets originating from c-quarks, hadronic decays of τ leptons and light jets respectively.

Re-clustered jets   Baseline jets that survive the overlap removal are re-clustered with the anti-k_T algorithm into large-R jets of radius $R = 0.8$, as described in Sect. 5.2.6. Re-clustered jets are trimmed removing the constituents whose p_T falls below 10% of the p_T of the re-clustered jet, and are then required to have $p_T > 100$ GeV and $|\eta| < 2.0$.

Electrons   Baseline electrons have to satisfy the Loose identification OP (see Sect. 5.5.2) and required to have $p_T > 20(5)$ GeV the analysis discussed in Chaps. 8, 9 and $|\eta| < 2.47$. After overlap removal, signal electrons are required to satisfy the Tight identification OP and the *LooseTrackOnly* isolation OP, and to have $p_T > 20$ GeV. Electrons are matched to the PV by requiring $|d_0/\sigma_{d_0}| < 5$ and $|z_0 \sin\theta| < 0.5$ mm.

Muons   Baseline muons are required to satisfy the Medium identification OP (see Sect. 5.4.2) and to have $p_T > 20(5)$ GeV in the analysis discussed in Chaps. 8, 9 and $|\eta| < 2.5$. After overlap removal, signal muons are required to satisfy the *LooseTrackOnly* isolation OP and to have $p_T > 20$ GeV. Muons are also matched to the PV by requiring $|d_0/\sigma_{d_0}| < 3$ and $|z_0 \sin\theta| < 0.5$ mm; muons that do not satisfy this selection are considered as cosmic muons, and the events with at least one cosmic muons are vetoed.

The missing transverse momentum is defined considering all the calibrated objects in the event, as described in Sect. 5.6, and uses the TST to account for the contribution of detector signals not associated to reconstructed objects.

7.4.1 Overlap Removal

Potential overlaps between jets, electrons and muons are resolved sequentially with the overlap removal procedure. First of all, electrons that are likely to originate from muon bremsstrahlung, i.e. those that lie within $\Delta R < 0.01$ from a muon candidate, are removed.

Overlaps between jet candidates and electron arise mostly from two reasons. Electrons are reconstructed from deposits in the calorimeter, and are therefore reconstructed as jets as well. In this case we want to preserve the electron and remove the jet, so we remove any non-b-tagged jet that lies within $\Delta R < 0.2$ from an electron candidate; an exception is done for b-tagged jets, since in this case the electron is likely to originate from a semileptonic B-hadron decay. Electron candidates that are close to the surviving jets are likely to have been produced in the decay of a hadron inside the jet, and are therefore removed if they lie within $\Delta R < 0.4$ from a jet and their energy is lower than 50 GeV, or if their energy is higher than 50 GeV and the ΔR from the jet is lower than $\min(0.4, 0.04 + 10\,\text{GeV}/E_T)$. Having an energy-dependent ΔR selection aims at increasing the acceptance for leptons originating from the decay of boosted top quarks.

Unlike electrons, muons are unlikely to be reconstructed as jets. Therefore, the first step in the overlap removal procedure between muon candidates and jets aims primarily at removing the muons that originate from the decay of hadrons inside the jets. Muons and jets can also be close if the jet is originating from muon bremsstrahlung; these jets typically have a small number of associated ID tracks. Therefore jets that lie within $\Delta R < 0.2$ from a muon are removed if they are not b-tagged and if they have less than three matching ID tracks. Muons in close proximity to the surviving jets are removed if they lie within $\Delta R < 0.4$ from a jet and their p_T is lower than 50 GeV, or if their p_T is higher than 50 GeV and the ΔR from the jet is lower than $\min(0.4, 0.04 + 10\,\text{GeV}/p_T)$.

7.5 Experimental Systematic Uncertainties

Each of the selections in the definition of the objects described in Sect. 7.4 has an associated experimental systematic uncertainty due to its modeling in the simulation. The experimental systematic uncertainties that are most relevant for analyses targeting final states with multiple b-jets and E_T^{miss} are:

JES and JER   As discussed in Sect. 5.2.3, the JES and JER calibration procedures allow to both calibrate the jet energy from the EM to the hadronic scale, and also to correct MC simulations to describe better the data; the uncertainties on these corrections propagate to the analyses. For the JES, reduced sets of uncertainties are available that reduce the number of NPs to be included in the analysis at the cost of potential loss in correlation; the analyses discussed in this thesis use a strongly reduced set with three NPs. Uncertainties on JES and JER are the leading source of experimental uncertainties, as they can change the kinematics of the jets in the event, and therefore also the number of jets that fulfill certain selections.

b-**tagging**   The uncertainties on the b-tagging calibration procedure (see Sect. 5.3.1) are included through uncertainties on the b-tagging SF. They are divided into one NP for the uncertainty on the efficiency of tagging jets originating from b-quarks, one for the mistagging of the jets originating from c-quarks, and one for light jets; furthermore, one additional NP for the high-p_{T} extrapolation uncertainty, based on MC simulation studies, increases the uncertainties in the high-p_{T} region, where the b-tagging calibration is not available and the central value of the SF is taken from the closest calibrated p_{T} region. The uncertainties on b-tagging play a relevant role in the analyses in this thesis, as they rely heavily on the selection of events containing b-tagged jets.

Luminosity   Besides those uncertainties associated with the object selections, another experimental uncertainty that is taken into account in the analysis is the uncertainty on the integrated luminosity, whose measurement is described in Sect. 3.3.6; for the 2015–2016 dataset the luminosity uncertainty corresponds to 3.2% and is derived with a strategy similar to that described in Ref. [3]. The impact of this uncertainty on the analyses discussed in this thesis is very small.

The following systematic uncertainties have been explicitly tested to be negligible in the analyses and are therefore not included in the final results presented in this thesis:

Leptons   The uncertainties on the calibration and resolution of the energy (momentum) of electrons (muons).

$E_{\mathrm{T}}^{\mathrm{miss}}$ The three NPs that take into account the uncertainty on the scale and the parallel and perpendicular resolution of the $E_{\mathrm{T}}^{\mathrm{miss}}$ soft term. Note that, while the $E_{\mathrm{T}}^{\mathrm{miss}}$ soft term uncertainties are not included in the final results, the propagation of the JES and JER uncertainties to the $E_{\mathrm{T}}^{\mathrm{miss}}$ computation are always properly taken into account.

7.6   Common Kinematic Variables

The kinematic properties of the objects in the event are used to define variables that, by combining information about multiple objects, provide a handle to discriminate between the signal events from the benchmark model being tested, and the background events from SM processes, described in Sect. 7.7. While a few of these variables are related to specific characteristics of the signal model, especially in the case of the higgsino search discussed in Chap. 9 that involves two Higgs bosons, most of them are built based on the features of the main background processes we need to suppress and are therefore in common between the two analyses.

Object energy and multiplicity Even just the multiplicity of the objects defined in Sect. 7.4 provides a powerful handle to analyze the characteristics of an event. In particular, the following multiplicity variables are used:

N_{jet} Number of signal jets in the event.
N_{b-jet} Number of signal jets that are b-tagged using the 77% OP.
N_{lepton} Number of signal leptons (electrons and muons) in the event.

In addition to multiplicity variables, we can also use the energy of the individual objects as a discriminating variable; for example, we always consider events with a minimum E_T^{miss} of 200 GeV (the selection is tighter in some SRs), or we can consider only jets with p_T above a certain threshold if we expect high-p_T jets from the signal.

Effective mass The effective mass is defined as the scalar sum of the momenta of all the signal objects in the event and the missing transverse momentum:

$$m_{eff}^{incl} = \sum_{i \leq n} p_T^{j_i} + \sum_{j \leq m} p_T^{\ell_j} + E_T^{miss} .$$

This variable reflects the overall energy scale of the event, and in the signal is therefore correlated to the mass of the SUSY particles produced.

Transverse mass The transverse mass between lepton and E_T^{miss} (m_T) is defined for events with at least one selected signal lepton as:

$$m_T = \sqrt{2 p_T^{lep} E_T^{miss} (1 - \cos \Delta\phi(E_T^{miss}, p_T^{lep}))} ,$$

where p_T^{lep} is the transverse momentum of the leading signal lepton in the event. For events where this lepton and the entire E_T^{miss} derive from the decay of a parent particle with mass m_{parent}, m_T presents an endpoint at m_{parent}. This is the case when the lepton and E_T^{miss} derive from the leptonic decay of a W boson, in W+jets events or in semileptonic $t\bar{t}$ events; note that dileptonic $t\bar{t}$ events do not present the same endpoint at the W-boson mass, as it is not possible to completely disentangle the energy of the two neutrinos. Events from W+jets or semileptonic $t\bar{t}$ processes can have values of m_T higher than the kinematic endpoint only if some of the objects in the event are not properly reconstructed (e.g. fake E_T^{miss} from mismeasured jets, jets misidentified as leptons, or leptons out of acceptance).

Another useful variable to suppress the $t\bar{t}$ background is the minimum transverse mass between E_T^{miss} and the b-jets in the event:

$$m_{T,min}^{b\text{-jets}} = \min_{i \leq 3}\sqrt{(E_T^{miss} + p_T^{j_i})^2 - (E_{Tx}^{miss} + p_x^{j_i})^2 - (E_{Ty}^{miss} + p_y^{j_i})^2} \; ,$$

where the minimum is taken over the transverse mass computed with the three leading b-tagged jets in the event. In semileptonic $t\bar{t}$ events, this variable presents an endpoint near $\sqrt{m(\text{top})^2 - m(W)^2}$, as this is the endpoint that is obtained when the E_T^{miss} originating from the neutrino is perfectly measured and the b-jet produced in the decay of the same top quark as the neutrino is among the three leading b-tagged jets.

Multijet suppression   SM background events without any neutrino in the decay chain can still have a sizable amount of reconstructed E_T^{miss} if one of the jets in the event is mismeasured, leading to a fake energy imbalance in the event. In these cases, the resulting E_T^{miss} will have a value of the azimuthal angle (ϕ) close to the one of the mismeasured jet. The variable $\Delta\phi_{min}^{4j}$ is defined as the minimum difference in azimuthal angle between E_T^{miss} and the four jets with the highest p_T in the event:

$$\Delta\phi_{min}^{4j} = \min_{i \leq 4}(|\phi_{\text{jet},i} - \phi_{E_T^{miss}}|) \; .$$

Requiring high values of this variable helps rejecting events with no real E_T^{miss}.

7.7   Background Processes and Their Modeling

This section describes the main background processes in the analysis regions, which are characterized by the selection of events with an high number of b-jets and high E_T^{miss}, and the way they are modeled. In general, the modeling is based on MC simulations for all the backgrounds, except multijet which is estimated with a data-driven technique. For the main background, pair production of top quark pairs ($t\bar{t}$), the shape of the different distributions is obtained from MC but the normalization is data-driven, derived in specifically designed CRs and tested in VRs as described in Sect. 7.3.

7.7.1   Top Quark Pair Production

Given the presence of several b-jets in the final state, $t\bar{t}$ production in association with jets constitutes the main source of SM background in all the analysis regions. $t\bar{t}$ production, which at the LHC at 13 TeV has a cross-section of $831.8^{+19.8+35.1}_{-29.2-35.1}$ pb [4], is mediated by the strong interaction, and it can occur through quark-antiquark annihilation (Fig. 7.2a) or through gluon-gluon fusion (Fig. 7.2b–d). When the LHC is running at 13 TeV, the threshold fraction of the proton energy that must be carried by each parton in order to have sufficient energy to produce a system of two top quarks is about 2.6%. At these low fractions, the PDF of the gluon is higher than

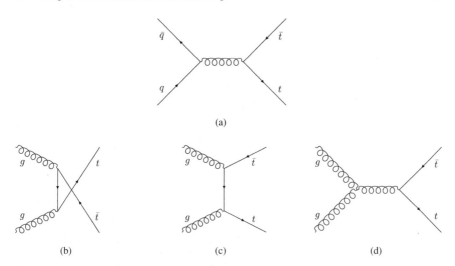

Fig. 7.2 LO Feynman diagram for the production of top quark pairs initiated by **a** quark-antiquark annihilation and **b–d** gluon-gluon fusion

the PDF of quarks and much higher than the PDF of the antiquarks, so gluon-gluon fusion is the dominant $t\bar{t}$ production mode at the LHC.

In the SM, the top quark decays $\approx 99.8\%$ of the times to a W boson and a b-quark. Beside the pair of b-quarks, which are always present, the final state can be characterized based on the decay of the two W bosons, leading to three different categories:

Dilepton   Both W bosons decay to a lepton (e, μ, τ) and a neutrino (10.5%).

Single-lepton   One W boson decays to lepton and neutrino, the other decays hadronically (43.8%).

All-hadronic   Both W bosons decay hadronically (45.7%).

All the analysis regions considered in this thesis have a tight E_T^{miss} selection (the loosest one is $E_T^{miss} > 200$ GeV). Since the all-hadronic component of the $t\bar{t}$ background does not have any neutrino in the final state, it does not fulfill the E_T^{miss} requirement unless one or more jets are mismeasured; this case produces a negligible background in the analysis regions, and is estimated with the jet-smearing method (described in Sect. 7.7.6.1) together with the other processes that pass the selection due to mismeasured jets. The $t\bar{t}$ components that are dominating the background estimate are the dilepton and single-lepton ones, both in the analysis regions that require reconstructed leptons as well as in the analysis regions that have a lepton veto. In the second case, the dominating component is the semi-leptonic $t\bar{t}$, where the lepton is a hadronically decaying τ, has a p_T too low to be reconstructed, or falls out of acceptance. The MC generator used to simulate $t\bar{t}$ events produced in association with high-p_T jets is POWHEG- BOX v2 with the CT10 PDF set [5] and interfaced with PYTHIA v6.428 [6] for the PS and hadronization. The normalization is

Fig. 7.3 $t\bar{t}$ decay type as a function of the m_T value in a selection requiring at least four jets, at least three *b*-jets, $E_T^{miss} > 200$ GeV and at least one signal lepton

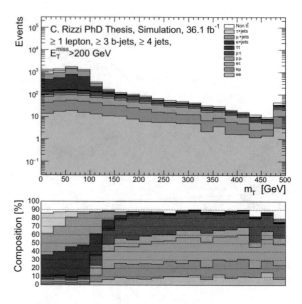

derived in CRs designed to be as kinematically close as possible to the corresponding SRs (while being non-overlapping with them by inverting some of the selections), and is expressed in terms of SFs relative to the cross-section computed with the highest available accuracy, which is NLO + next-to-next-to-leading logarithmic (NNLL) [7].

7.7.1.1   Truth-Level Classification: $t\bar{t}$ Decays

When considering events at particle level, the MC information can be used to reconstruct the decay chains that lead to the stable particles in the final state. Starting from each top quark in the event, it is possible to follow its decay chain and classify the event first of all into dilepton or single-lepton, or also with a finer classification based on the lepton flavor. Subsequent decays of the τ leptons are not considered, and hadronically decaying τ leptons are fully considered as leptons in this classification. Figure 7.3 shows the $t\bar{t}$ decay type as a function of the m_T value in a selection requiring at least four jets, at least three *b*-jets, $E_T^{miss} > 200$ GeV and at least one signal lepton. It is possible to see that, while at low m_T values the $t\bar{t}$ component is dominated by events where one top quark decays leptonically and the other one decays hadronically, after the kinematic threshold imposed by the mass of the W boson the dilepton component is dominant.

Figure 7.4 shows the same $t\bar{t}$ classification but as a function of $m_{T,min}^{b\text{-}jets}$, both in a selection with at least one signal lepton (Fig. 7.4a) and with a lepton veto (Fig. 7.4b). Here it is possible to notice two features. First of all, in general both the selection with a lepton requirement and the one with a lepton veto are dominated by semileptonic decays: in the case of the selection with the lepton requirement the lepton is in most

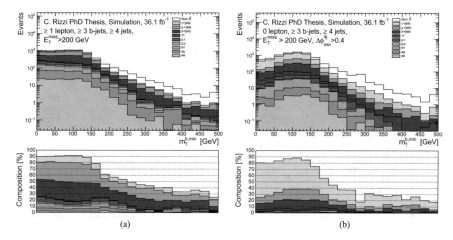

Fig. 7.4 $t\bar{t}$ decay type as a function of the $m_{\mathrm{T,min}}^{b\text{-jets}}$ value in a selection requiring at least four jets, at least three b-jets, $E_{\mathrm{T}}^{\mathrm{miss}} > 200$ GeV and **a** at least one signal lepton **b** exactly zero leptons and $\Delta\phi_{\mathrm{min}}^{4j} > 0.4$

of the cases a muon or an electron, while in the case of the selection with the lepton veto most of the events contain one τ lepton, which can decay hadronically giving rise to a topology with no electrons or muons. Secondly, while in the case of m_{T} $t\bar{t}$ events are dominating the background composition also for high values of this variable (just with a change from semileptonic to dileptonic $t\bar{t}$ events), in the case of $m_{\mathrm{T,min}}^{b\text{-jets}}$ after the kinematic endpoint the $t\bar{t}$ fraction decreases noticeably.

7.7.1.2 Truth-Level Classification: Flavor of the Associated Jets

A truth-level classification is in place to study the $t\bar{t}$ +jets background based on the flavor of the associated jets. This is done starting from the MC information at particle level; stable particles are grouped into jets using the anti-k_t algorithm described in Sect. 5.2.2 with R = 0.4; only particle jets with $p_{\mathrm{T}} > 15$ GeV and $|\eta| < 2.5$ are considered. The flavor of the jets is determined by matching them with the B- and D-hadrons that are in a cone of $\Delta R = 0.4$ from the jet. If the event contains at least one jet matched to one or more B-hadrons, excluding the ones originating from the decay of the top quarks, the event is classified as $t\bar{t} + \geq 1b$. Otherwise, if at least one jet is matched to D-hadrons, excluding the ones from the decay chain of the top quarks, the event is classified as $t\bar{t} + \geq 1c$. $t\bar{t} + \geq 1b$ and $t\bar{t} + \geq 1c$ events are categorized together as $t\bar{t}$ + HF events, where HF stands for heavy flavor, while the remaining ones are classified as $t\bar{t}$ + light−jets. Figure 7.5 shows the evolution of the flavor composition of the jets produced in association with $t\bar{t}$ as a function of the number of b-jets in the event. As expected, as the number of b-jets increases, the $t\bar{t}$ + light−jets fraction decreases and the $t\bar{t} + \geq 1b$ fraction increases.

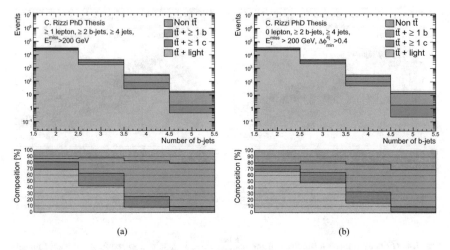

Fig. 7.5 Evolution of the flavor of the jets produced in association with $t\bar{t}$ in a selection requiring at least four jets, at least two b-jets, $E_T^{miss} > 200$ GeV and **a** at least one signal lepton **b** exactly zero leptons and $\Delta\phi_{min}^{4j} > 0.4$

The production of $t\bar{t}$ + HF is a process that has large theoretical uncertainties, and therefore in the searches discussed in this document it has assigned a 30% normalization uncertainty, in accordance with the results of the ATLAS measurement of this cross-section at $\sqrt{s} = 8$ TeV [8].

7.7.1.3   Modeling Uncertainties

The modeling uncertainties on the $t\bar{t}$ background considered in this thesis are:

Generator   The uncertainty associated with the choice of a specific MC generator is estimated by comparing POWHEG- BOX with MADGRAPH5_AMC@NLO, both interfaced with HERWIG++ v2.7.1 with the UEEE5 underlying-event tune.

Parton shower and hadronization   Also the choice of the generator that emulates the PS and the hadronization is associated to an uncertainty, that is evaluated by comparing the nominal sample, generated with POWHEG- BOX and showered with PYTHIA, to another sample generated again with POWHEG- BOX but showered with HERWIG++ v2.7.1.

Radiation   The systematic uncertainty related to the modeling of the ISR and FSR is estimated by comparing samples generated with POWHEG- BOX interfaced with two versions of PYTHIA v6.428 with two different settings [9]. One uses the PERUGIA2012radHi tune, has the h_{damp} parameter set to twice the top mass and the renormalization and factorization scales set to twice the nominal value; these settings lead to an overall larger amount of radiation. The second sample uses a version of PYTHIA with the PERUGIA2012radLo tune, has h_{damp} set to the top

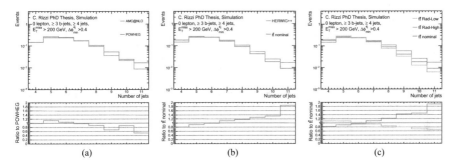

Fig. 7.6 Distribution normalized to unity of the number of signal jets in the $t\bar{t}$ MC sample in a selection requiring at least four jets, at least two b-jets, $E_T^{miss} > 200$ GeV and exactly zero leptons. **a** POWHEG- BOX (pink line) and MADGRAPH5_AMC@NLO (blue line), both interfaced with HERWIG++. **b** Nominal sample (pink line), generated with POWHEG- BOX and showered with PYTHIA, and a sample generated with POWHEG- BOX and showered with HERWIG++ (blue line). **c** Nominal sample (pink line) and two varied samples generated with POWHEG- BOX interfaced with two versions of PYTHIA (blue and green line)

mass and the renormalization and factorization scales set to half of the nominal value, leading to a description of the event with less additional jets.

Figures 7.6 and 7.7 show the changes in the shape of the distribution of the number of signal jets and the p_T of the leading jet when comparing these systematic variations in a representative selection with a lepton veto. The normalization of $t\bar{t}$ events is derived in the CRs, therefore the modeling uncertainties affect only the extrapolation from the CR to the corresponding SRs and VRs, and not the overall normalization. The modeling uncertainties are therefore estimated by comparing the expected values for the TFs, defined in Sect. 6.4.2 as the ratio of expected yields in the SR or VR over the expected yields in the corresponding CR. The uncertainties on the TFs obtained from the different sources listed above are summed in quadrature in each region, and are treated as uncorrelated across regions to avoid constraints from the fit.

The main limitation for the estimate of the uncertainties is the statistical uncertainty of the MC samples for the systematic variations, as these samples have in general less simulated events than the nominal sample. A higher number of simulated events is available at particle level, where the simulation of the interaction of the particles with the detector and the following event reconstruction (discussed in Sect. 4.4) is not performed, so the evaluation of the $t\bar{t}$ modeling uncertainties is carried out comparing samples at particle level, with the assumption that the detector simulation has a similar effect on all the samples, independently of the MC generator used. Two further strategies are in place to reduce the statistical uncertainty in the comparison of the different samples. One consists in relaxing or removing some selections in both the CRs and the corresponding SRs/VRs. Another is to substitute the selection on the number of b-jets with truth-tagging, described in the next section.

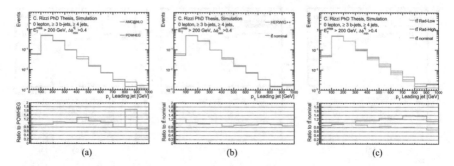

Fig. 7.7 Distribution normalized to unity of the p_T of the leading signal jet in the $t\bar{t}$ MC sample in a selection requiring at least four jets, at least two *b*-jets, $E_T^{miss} > 200$ GeV and exactly zero leptons. **a** POWHEG-BOX (pink line) and MADGRAPH5_AMC@NLO (blue line), both interfaced with HERWIG++. **b** nominal sample (pink line), generated with POWHEG-BOX and showered with PYTHIA, and a sample generated with POWHEG-BOX and showered with HERWIG++ (blue line). **c** Nominal sample (pink line) and two varied samples generated with POWHEG-BOX interfaced with two versions of PYTHIA (blue and green line)

7.7.1.4 Truth *b*-tagging

With truth tagging, instead of keeping only the events that satisfy a certain criterion on the number of *b*-tagged jets, all the events are kept and weighted. The weight is the probability that, out of all selected jets in the event, a certain number pass the *b*-tagging identification. A different weight is computed for each *b*-tagging requirement; for example, the probability for the event to have at least three *b*-tagged jets will be different from the probability of having exactly two. For each jet, the probability to be *b*-tagged can be expressed as a function of the jet flavor (f), p_T and η:

$$\varepsilon\left(f, |\eta|, p_T\right) . \tag{7.1}$$

If an event has N jets, the probability of containing exactly one *b*-tag jet can be expressed as:

$$P_{=1} = \sum_{i=1}^{N}\left(\varepsilon_i \prod_{i \neq j}\left(1 - \varepsilon_j\right)\right) ,$$

and in the same way, we can compute the probability for an inclusive *b*-tagging selection:

$$P_{=0} = \prod_{i=1}^{N}\left(1 - \varepsilon_j\right) ,$$

$$P_{\geq 1} = 1 - P_{=0} .$$

Fig. 7.8 Example scheme to choose the selected permutation with truth b-tagging

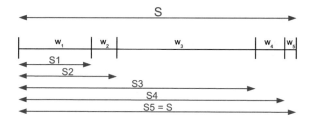

This procedure can be extended to an arbitrary number of b-tagged jets by summing over all the possible permutations that lead to the desired number of b-tagged jets (n) to derive the exclusive probability and then subtract it from the $P_{\geq n}$ probability to have the inclusive probability for $\geq n + 1$ b-tagged jets.

Beside decreasing the statistical uncertainty on the expected number of events, the other advantage of using truth tagging on particle-level samples is that it allows to emulate the reconstruction-level efficiency of the b-tagging algorithm: at particle-level a jet is considered to be b-tagged if it is matched to a b-quark in the MC record within a certain cone. This criterion has almost 100% efficiency for real b-jets, and it leads to a null mistag rate. Instead with truth tagging, if the efficiency map used in Eq. 7.1 corresponds to the b-tagging efficiency measured for reconstructed jets, it is possible to obtain the same efficiency to a b-tagging selection as in a reconstructed-level analysis.

Some analysis variables are built using explicitly the kinematic characteristics of the b-tagged jets, e.g. $m_{\mathrm{T,min}}^{b\text{-jets}}$. The truth tagging method allows to choose which jets in the event should be considered as b-tagged: in an event with N jets, each permutation with n jets considered as b-tagged and $N - n$ jets considered as non b-tagged is characterized by an individual weight w_i. If we let S to be the sum of all these individual weights, a pseudo-random number generated with uniform probability between 0 and S will indicate which permutation to choose and consequently which jets to consider as b-tagged; this procedure is illustrated in Fig. 7.8.

7.7.2 Single Top Quark Production

While the production of a $t\bar{t}$ pair is mediated by the strong interaction, in the case of a single top quark the production occurs via the electroweak interaction. There are three possible production channels, illustrated in Fig. 7.9: the t-channel, the Wt-channel, where a top quark and a W boson are produced, and the s-channel. The Wt- and s-channel production mechanisms are simulated with the same generator choice as $t\bar{t}$: POWHEG- BOX v2, showered with PYTHIA v6.428. Single-top events produced through the t-channel process are simulated with POWHEG- BOX v1, which uses the four-flavor scheme for the computation of the NLO ME and the CT10f4 PDF set (with the four-flavor scheme as well); the top quarks are decayed with MADSPIN [10]. The

Fig. 7.9 LO Feynman diagrams for single-top production: **a–b** Wt-channel, **b** s-channel, **d** t-channel

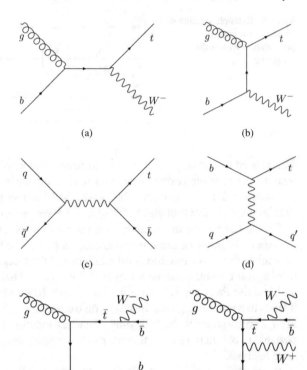

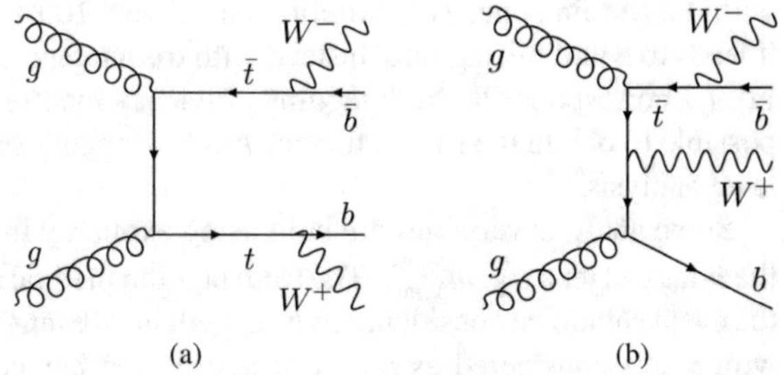

(a)   (b)

(c)   (d)

Fig. 7.10 Two example diagrams that lead to a $WWbb$ final state. **a** double resonance, both Wb pairs form a top quark. **b** single resonance, only one Wb pair forms a top quark

(a)   (b)

cross-section used for the normalization of the single-top processes is computed at NLO+NNLL order [11–13].

7.7.2.1   Wt-Channel and $t\bar{t}$ Interference

While the LO diagrams for Wt-channel single-top production in Fig. 7.9a and b show a clear signature characterized by one single heavy-flavor quark in the final state, when NLO diagrams are considered it is possible to reach the same final state as the $t\bar{t}$ process, namely $WWbb$, as shown in Fig. 7.10. Considering the two processes separately leads necessarily to an improper treatment of the quantum interference between them. POWHEG- BOX allows to approach this problem with two different strategies, diagram removal (DR) and diagram subtraction (DS) [14]; while none of them reproduces the quantum interference between single top and $t\bar{t}$, they allow to mitigate the size of the effect.

We can write the matrix element for the production of the $WWbb$ final state as:

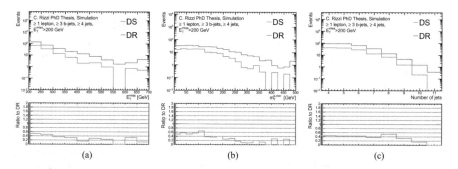

Fig. 7.11 Distribution of **a** E_T^{miss}, **b** $m_{T,min}^{b\text{-jets}}$ and **c** number of signal jets in the single-top MC sample obtained with the DR and DS approaches (pink and blue line respectively), in a selection requiring at least four jets, at least two b-jets, $E_T^{miss} > 200$ GeV and at least one lepton

$$\mathcal{M}_{WWbb} = \mathcal{M}_{\text{double-res}} + \mathcal{M}_{\text{single-res}} ,$$

where the first term represents the contribution from $t\bar{t}$ production (double resonance) and the second term the contribution from NLO corrections to the Wt single-top production, where only one of the two Wb systems is resonant (single resonance). When computing the squared amplitude, this becomes:

$$|\mathcal{M}_{WWbb}|^2 = |\mathcal{M}_{\text{double-res}}|^2 + |\mathcal{M}_{\text{single-res}}|^2 + 2\mathcal{R}\left(\mathcal{M}_{\text{double-res}} \times \mathcal{M}_{\text{single-res}}^*\right) .$$

While the first two terms are properly taken into account by the individual $t\bar{t}$ and single-top MC simulations, the last term represents the interference between the two processes and, unless special care to avoid this is taken, it is completely ignored if the two processes are simulated independently.

In the DR approach, all the diagrams where a W boson and a b-quark form a top quark are removed from the matrix element used to compute the single-top cross-section in the Wt channel. While this approach is not gauge invariant, it is found to have negligible dependence on the gauge choice.

Instead in the DS approach an extra subtraction term is added directly to the differential cross-section to cancel the contribution of the production of two on-shell top quarks, which is more accurate the more the invariant mass of the non-resonant Wb system tends to the top quark mass. Acting at the cross-section level and not at the matrix-element level, this approach aims at accounting also for the interference term.

Which one of the two strategies gives a better description of the data depends on the phase space under consideration. In this thesis the nominal single-top estimate for the Wt channel is obtained with the DR strategy. As an example, Fig. 7.11 shows how the choice of the DR or DS approach can change the distribution of two key variables in the analyses, E_T^{miss} and $m_{T,min}^{b\text{-jets}}$.

To evaluate the uncertainty associated with the strategy chosen to account for the interference between $t\bar{t}$ and single top we use a dedicated $WWbb$ sample generated at particle level with MADGRAPH5_AMC@NLO showered with PYTHIA v8, that takes fully into account the interference. This sample is produced at LO and using the four-flavor scheme, where b-quarks are treated as massive and processes that require a b-quark in the initial state (which is the case for Wt production) are initiated by a gluon splitting into a $b\bar{b}$ pair. Because of the LO accuracy, the comparison with the sum of the two nominal samples for $t\bar{t}$ and Wt, generated with POWHEG- BOX and showered with PYTHIA, does not lead to sensible results as the NLO corrections to the $t\bar{t}$ sample are larger than the effect of the interference. Instead, two additional samples are generated with the same settings as the $WWbb$ sample to simulate the double-resonance and the single-resonance samples: the former requires the presence of two resonant top quarks, the latter requires one resonant top quark and two b-quarks in the ME, vetoing the presence of a second resonant top quark.

7.7.2.2 Modeling Uncertainties

The modeling uncertainties on the single-top background are estimated at particle level, like in the case of the uncertainties on the $t\bar{t}$ background, but in this case the uncertainty is computed based on the difference of the expected yields in SRs and VRs and not based on the difference between the TFs; this is because in the analyses discussed in this thesis the single-top background is normalized to its theoretical cross-section and does not have a data-driven normalization. Truth tagging is also used in this case to increase the available number of simulated events, and to provide a description of the b-tagging efficiency closer to the one that we have in the reconstructed events. The modeling uncertainties considered are:

Interference   The uncertainty deriving from the treatment of the interference between $t\bar{t}$ and single top in the Wt channel is obtained by comparing the sum of single-resonance and double-resonance contributions to the total $WWbb$ sample, using the samples described in the previous paragraph.
Radiation   The systematic uncertainty on the modeling of the extra radiation is estimated with dedicated samples generated with the same set of varied parameters used to generate the variation samples for $t\bar{t}$, described in Sect. 7.7.1.

7.7.3 *Vector Boson in Association with Jets*

The production of a vector boson (W or Z boson) in association with jets is modeled in ATLAS with the SHERPA generator in version 2.2: MEs are computed with COMIX [15] and OPENLOOPS [16], and merged with the SHERPA parton shower with the CKKW prescription; the PDF set used is NNPDF 3.0, and the expected number of events is normalized to the next-to-next-to-leading order (NNLO) cross-section [17]. As an

Fig. 7.12 Representative
LO Feynman diagrams for
the production of a W boson
in association with one jet

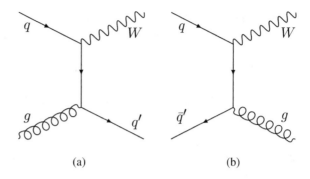

(a)   (b)

illustration, the LO Feynman diagrams for the production of a W boson in association
with one jet are shown in Fig. 7.12. The diagrams for the production of a Z boson in
association with one jet are the same, except that in the case of the Z boson the two
quarks in Fig. 7.12a and the quark and antiquark in Fig. 7.12b have the same flavor.

Our analysis selections require high E_T^{miss} and the presence of b-jets, therefore
most of the W+jets events that fulfill the selection are the ones where the W boson
decays to an electron, muon or τ lepton and the associated neutrino (to produce
some E_T^{miss}) and the b-tagging requirement is satisfied because of the production of
extra heavy-flavor jets, or because the jet from the hadronic decay of the τ lepton
is incorrectly tagged as a b-jet together with other associated jets. W+jets events
can therefore be present both in analysis regions with a lepton requirement and in
those with a veto on the presence of leptons. In the case of Z+jets events, most of
the events that enter the analysis regions have a $Z \rightarrow \nu\nu$ decay (which has a BR of
about 20%), and are produced in association with heavy-flavor jets or jets that are
mistagged; these events contribute almost exclusively to the analysis regions with a
lepton veto.

7.7.3.1 Modeling Uncertainties

The modeling uncertainties considered in the case of the simulation of vector boson
production in association with jets are related to the choice of parameters in the
SHERPA generator. In particular, the parameters whose choice can influence the
description of the events are:

Renormalization scale   The default renormalization scale (μ_R) is set to the mass
of the W or Z boson in the W+jets and Z+jets samples respectively. The systematic
uncertainty associated with this choice is derived by comparison with alternative
samples where μ_R is set to twice or to half of the nominal value.

Factorization scale   Like μ_R, the factorization scale (μ_F) has its default value at
the mass of the vector boson, and is varied to twice and half this value.

Matching scale   The scale for the matching between ME and PS is set to 20 GeV, and this choice is compared with two alternative samples where it is set to 30 and 15 GeV.

Resummation scale   As in the case of μ_{R} and μ_{F}, the central value of the resummation scale (Q_{sf}, the scale used to factorize between constant and logarithmic terms in the resummation process) is set to the W or Z boson mass, and varied to twice or half its central value in the systematic uncertainties.

In order to derive the effect of these systematic uncertainties, the analyses described in this thesis do not use a direct comparison of the nominal samples with the ones implementing the systematic variations. Instead, the comparison is implemented through weights with a 2D parametrization based on the number of jets in the event at particle level and on the transverse momentum of the vector boson [18]. As an example, the impact of these systematic uncertainties on the shapes of the distributions of $E_{\mathrm{T}}^{\mathrm{miss}}$ in a selection with a lepton veto is shown in Fig. 7.13 for the Z+jets sample and in Fig. 7.14 for the W+jets sample.

7.7.4 Diboson Production

In addition to the production of a single vector boson, also pair production is possible: the LO Feynman diagrams for the production of a boson pair are shown in Fig. 7.15. The cross-section is suppressed by more than three orders of magnitude with respect to the single-vector-boson case. Therefore, despite the fact that presence of two bosons leads a higher acceptance into the analysis regions, the overall contribution of diboson processes to the total SM background is small. The diboson MC samples are generated with SHERPA v2.2.1, and their cross-sections are computed at NLO [19, 20]. A 50% normalization uncertainty is applied to the predicted number of events for this background, to take into account systematic uncertainties on its modeling. This conservative uncertainty is found to have a negligible impact on the analysis results.

7.7.5 $t\bar{t} + X$ Production

A pair of top quarks can be produced also in association with a vector boson o a Higgs boson, as shown in Fig. 7.16a–c. In this thesis, these processes constitute a minor background and are grouped in the category $t\bar{t} + X$, together with four-top production, for which an example LO Feynman diagram is shown in Fig. 7.16d. The production of a $t\bar{t}$ pair in association with a vector boson and the four-top production are modeled with MADGRAPH5_AMC@NLO showered with PYTHIA v8, while for $t\bar{t} + H$ the same ME generator is used but the PS processes is performed with HERWIG++. All of these samples are normalized to the corresponding NLO cross-

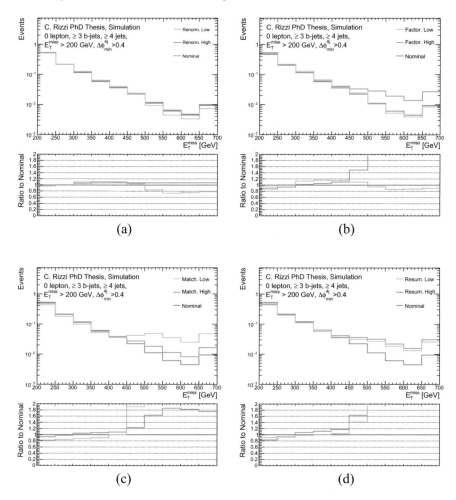

Fig. 7.13 Distribution normalized to unity of the p_T of the leading signal jet in the Z+jets MC sample in a selection requiring at least four jets, at least two b-jets, $E_T^{miss} > 200$ GeV and exactly zero leptons. **a** Comparison of nominal and renormalization scale variations. **b** Comparison of nominal and factorization scale variations. **c** Comparison of nominal and matching scale variations. **d** Comparison of nominal and resummation scale variations

sections [21, 22]. As described above for the diboson background, a conservative 50% normalization uncertainty is applied also to the $t\bar{t} + X$ background, and also in this case it is found to have negligible impact in the analysis.

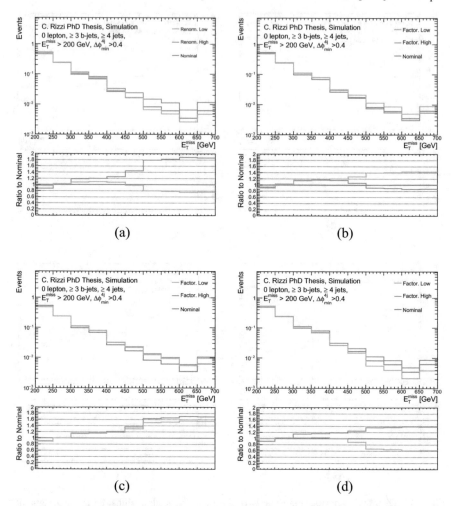

Fig. 7.14 Distribution normalized to unity of the p_T of the leading signal jet in the *W*+jets MC sample in a selection requiring at least four jets, at least two *b*-jets, $E_T^{miss} > 200$ GeV and exactly zero leptons. **a** Comparison of nominal and renormalization scale variations. **b** Comparison of nominal and factorization scale variations. **c** Comparison of nominal and matching scale variations. **d** Comparison of nominal and resummation scale variations

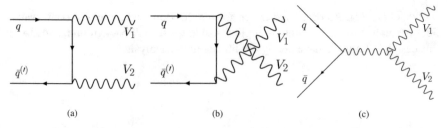

Fig. 7.15 Representative LO Feynman diagrams for diboson production ($V = W, Z$)

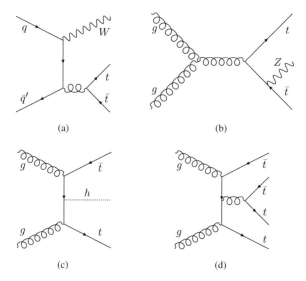

Fig. 7.16 Representative LO Feynman diagram for the production of **a** $t\bar{t}W$, **b** $t\bar{t}Z$, **c** $t\bar{t}H$, and **d** $t\bar{t}t\bar{t}$

7.7.6 QCD Multijet

Multijet production is by far the process with the largest cross-section at the LHC ($\mathcal{O}(\text{mb})$). Most of the QCD events are $2 \rightarrow 2$ processes, for which some example LO Feynman diagrams are shown in Fig. 7.17a and b, but higher order processes ($2 \rightarrow n$) are possible as well (an example of $2 \rightarrow 3$ process is shown in Fig. 7.17c). Multijet processes do not yield real E_T^{miss}, and therefore they can be a background in analyses that require high E_T^{miss} only if one or more of the jets are mismeasured, leading to an energy imbalance in the transverse plane that is reconstructed as E_T^{miss}. These processes do not produce any lepton either, so also a contribution to analyses requiring leptons must originate from a jet misreconstructed as a lepton. Since the probability for a multijet event to both have a fake high E_T^{miss} and a fake lepton is very small, multijet processes are considered as background only in the analysis regions with a veto on the presence of leptons.

7.7.6.1 Jet Smearing

The usage of MC simulation to model the multijet background presents two main drawbacks. First of all, the cross-section for multijet production is very difficult to predict accurately. It is feasible to model the $2 \rightarrow 2$ dijet production, but every additional parton in a $2 \rightarrow n$ process brings into the computation a further α_s factor, and the energy scale of the interaction is low enough to be at the limit of the validity of the perturbative expansion, limiting the validity of a LO order computation. The

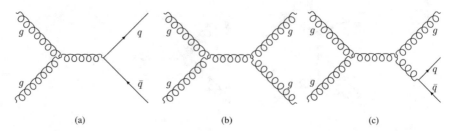

(a)   (b)   (c)

Fig. 7.17 Representative LO Feynman diagram for multijet production. **a–b** $2 \rightarrow 2$ process. **c** $2 \rightarrow 3$ process

inclusion of corrections of higher order can improve the description, but the number of extra diagrams that should be included makes this option unrealistic (just the LO computation for the $2 \rightarrow 2$ process comprises 10 different diagrams). A second problem is of practical nature: the high cross-section for multijet production implies a huge number of MC events that need to be simulated in order not to have a huge statistical uncertainty.

The jet smearing method overcomes these problems by providing a data-driven estimate of the multijet background, through the following steps:

- A sample of well-measured multijet seed events, with low values of E_T^{miss}, is selected in data.
- Each jet in each seed event is smeared: its four-momentum is multiplied by a random number found from sampling the jet response function, that quantifies the fluctuations in the p_T reconstruction of the jets. The response, defined as the ratio E_T^{reco}/E_T^{truth}, is determined from simulations and is then corrected to match data measurements. A different response function is used for b-tagged jets, since the neutrinos originating from the semileptonic decay of B-hadrons affect the response of the jet.
- The smearing procedure is repeated a large number of times, $\mathcal{O}(1000)$, for each event, and every time E_T^{miss} is recomputed with the smeared jets as input; some of these smeared events will have high E_T^{miss} values and will satisfy the analysis selections.
- The number of events that satisfy the analysis selections is arbitrary and depends on the choice of the seed events and on the number of smears for each event. The normalization is derived in a specifically designed QCD CR, that relies on a tight upper selection on $\Delta\phi_{min}^{4j}$ ($\Delta\phi_{min}^{4j} < 0.1$). All the 0-lepton analysis regions have the requirement $\Delta\phi_{min}^{4j} > 0.4$, which makes them orthogonal to the QCD CR and allows also to have VRs in the intermediate region.

Figure 7.18 shows the distribution of $\Delta\phi_{min}^{4j}$ after a selection requiring at least four jets, at least two b-jets, $E_T^{miss} > 200$ GeV and exactly zero leptons. The first bin of this distribution is the region used for the normalization of the multijet background, while the phase space up to $\Delta\phi_{min}^{4j} = 0.4$ (which is the lower threshold for the analysis regions) serves as validation region. Figures 7.19 and 7.20 show respectively the

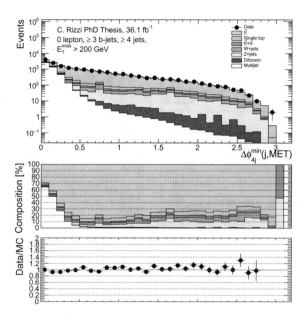

Fig. 7.18 Distribution of $\Delta\phi_{min}^{4j}$ in a selection requiring at least four jets, at least two b-jets, $E_T^{miss} > 200$ GeV and exactly zero leptons

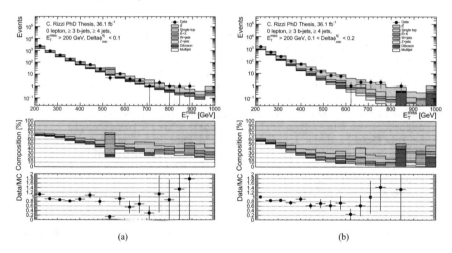

Fig. 7.19 Distribution of E_T^{miss} in **a** the region with $\Delta\phi_{min}^{4j} < 0.1$ where the data-driven estimate for the multijet background is normalized and **b** the multijet validation region with $0.1 < \Delta\phi_{min}^{4j} < 0.2$

distribution of E_T^{miss} and of the number of jets in the QCD CR and in a validation region with $0.1 < \Delta\phi_{min}^{4j} < 0.2$.

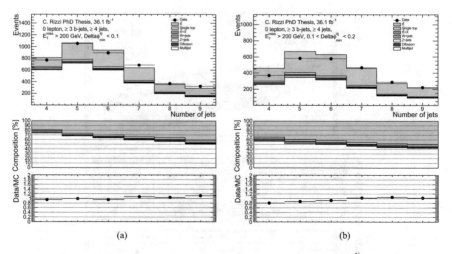

Fig. 7.20 Distribution of the number of signal jets in **a** the region with $\Delta\phi_\mathrm{min}^{4j} < 0.1$ where the data-driven estimate for the multijet background is normalized and **b** the multijet validation region with $0.1 < \Delta\phi_\mathrm{min}^{4j} < 0.2$

References

1. ATLAS Collaboration (2017) Performance of the ATLAS trigger system in 2015. Eur Phys J C 77:317. arXiv:1611.09661 [hep-ex]
2. LHC New Physics Working Group Collaboration, Alves D (2012) Simplified models for LHC new physics searches. J Phys G 39:105005. arXiv:1105.2838 [hep-ph]
3. ATLAS Collaboration (2016) Luminosity determination in pp collisions at $\sqrt{s} = 8$ TeV using the ATLAS detector at the LHC. Eur Phys J C 76:653. arXiv:1608.03953 [hep-ex]
4. Czakon M, Fiedler P, Mitov A (2013) Total top-quark pair-production cross section at hadron colliders through $O(\alpha_S^4)$. Phys Rev Lett 110:252004. arXiv:1303.6254 [hep-ph]
5. Lai H-L, Guzzi M, Huston J, Li Z, Nadolsky PM, Pumplin J, Yuan CP (2010) New parton distributions for collider physics. Phys Rev D 82:074024. arXiv:1007.2241 [hep-ph]
6. Sjostrand T, Mrenna S, Skands PZ (2006) PYTHIA 6.4 physics and manual. JHEP 05:026. arXiv:hep-ph/0603175 [hep-ph]
7. Czakon M, Mitov A (2014) Top++: a program for the calculation of the top-pair cross-section at hadron colliders. Comput Phys Commun 185:2930. arXiv:1112.5675 [hep-ph]
8. ATLAS Collaboration (2016) Measurements of fiducial cross-sections for $t\bar{t}$ production with one or two additional *b*-jets in *pp* collisions at $\sqrt{s} = 8$ TeV using the ATLAS detector. Eur Phys J C 76:11. arXiv:1508.06868 [hep-ex]
9. Skands PZ (2010) Tuning Monte Carlo generators: the perugia tunes. Phys Rev D 82:074018. arXiv:1005.3457 [hep-ph]
10. Artoisenet P, Frederix R, Mattelaer O, Rietkerk R (2013) Automatic spin-entangled decays of heavy resonances in Monte Carlo simulations. JHEP 03:015. arXiv:1212.3460 [hep-ph]
11. Kidonakis N (2011) Next-to-next-to-leading-order collinear and soft gluon corrections for t-channel single top quark production. Phys Rev D 83:091503. arXiv:1103.2792 [hep-ph]
12. Kidonakis N (2010) Two-loop soft anomalous dimensions for single top quark associated production with a W^- or H^-. Phys Rev D 82:054018. arXiv:1005.4451 [hep-ph]
13. Kidonakis N (2010) NNLL resummation for s-channel single top quark production. Phys Rev D 81:054028. arXiv:1001.5034 [hep-ph]

14. Frixione S, Laenen E, Motylinski P, Webber BR (2006) Single-top production in MC@NLO. JHEP 03:092. arXiv:hep-ph/0512250 [hep-ph]
15. Gleisberg T, Hoeche S (2008) Comix, a new matrix element generator. JHEP 12:039. arXiv:0808.3674 [hep-ph]
16. Cascioli F, Maierhofer P, Pozzorini S (2012) Scattering amplitudes with open loops. Phys Rev Lett 108: 111601. arXiv:1111.5206 [hep-ph]
17. Catani S, Cieri L, Ferrera G, de Florian D, Grazzini M (2009) Vector boson production at hadron colliders: a fully exclusive QCD calculation at NNLO. Phys Rev Lett 103:082001. arXiv:0903.2120 [hep-ph]
18. Anders JK, D'Onofrio M, Kretzschmar J (2017) Searches for direct pair production of third generation squarks, and dark matter, in final states containing $b-$ jets and E_{Tmiss} using the ATLAS detector at the LHC. https://cds.cern.ch/record/2291836. CERN-THESIS-2017-216
19. ATLAS Collaboration (2016) Multi-boson simulation for 13 TeV ATLAS analyses. ATL-PHYS-PUB-2016-002. https://cds.cern.ch/record/2119986
20. ATLAS Collaboration (2017) Multi-Boson simulation for 13 TeV ATLAS analyses. ATL-PHYS-PUB-2017-005. https://cds.cern.ch/record/2261933
21. Alwall J, Frederix R, Frixione S, Hirschi V, Maltoni F, Mattelaer O, Shao HS, Stelzer T, Torrielli P, Zaro M (2014) The automated computation of tree-level and next-to-leading order differential cross sections, and their matching to parton shower simulations. JHEP 07:079. arXiv:1405.0301 [hep-ph]
22. LHC Higgs Cross Section Working Group Collaboration, Andersen JR et al. Handbook of LHC Higgs cross sections: 3. Higgs properties. arXiv:1307.1347 [hep-ph]

Chapter 8
Search for Gluino Pair Production

This chapter presents a search for gluino pair production, leading to final states with multiple top and/or bottom quarks and E_T^{miss}. Section 8.1 describes the signal models used to optimize and interpret the analysis. Section 8.2 presents some kinematic variables specific to this analysis and illustrates how they can be used to discriminate signal from background. The analysis regions are defined in Sects. 8.3 and 8.4 for the cut-and-count and multi-bin strategies respectively.

Section 8.5 shows the comparison between data and simulation in a kinematic regime close to that of the analysis regions. Section 8.6 discusses the systematic uncertainties that are included in the analysis, and Sect. 8.7 presents the results. The interpretation of the results in terms of model-independent and model-independent limits is presented in Sect. 8.8. In Sect. 8.10 we show the update of the analysis including also the 2017 data taking period.

8.1 Signal Model

The simplified models used to optimize this analysis are the two gluino-pair-production models shown in Fig. 8.1. Figure 8.1a shows a schematic diagram of the Gtt model: each of the pair-produced gluinos decays with 100% BR to two top quarks and the lightest neutralino ($\tilde{\chi}_1^0$). In the Gbb model, shown in Fig. 8.1b, the decay happens through an off-shell sbottom and each gluino decays to two bottom quarks and the $\tilde{\chi}_1^0$. Both models assume R-parity conservation, so the $\tilde{\chi}_1^0$, which in this case is the LSP, is stable and escapes the detection giving rise to E_T^{miss} in the event. In both cases, the three body decay is realized through an off-shell squark, and it involves two interaction vertices. The first one is a strong-interaction vertex, in the case of the Gtt model: $\tilde{g} \rightarrow t\tilde{t}^{(*)}$. The second is an electroweak vertex, originating from the decay of the $\tilde{t}$ (or $\tilde{b}$ in the case of the Gbb model): $\tilde{t} \rightarrow t\tilde{\chi}_1^0$.

© Springer Nature Switzerland AG 2020
C. Rizzi, *Searches for Supersymmetric Particles in Final States with Multiple Top and Bottom Quarks with the Atlas Detector*, Springer Theses, https://doi.org/10.1007/978-3-030-52877-5_8

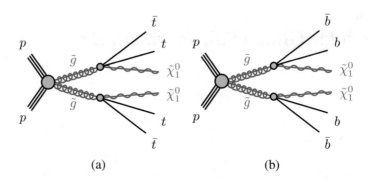

Fig. 8.1 The simplified models used for the optimization of the analysis. **a** Gtt model. **b** Gbb model

In the Gtt and Gbb models, since the $\tilde{t}$ and $\tilde{b}$ are assumed to have very high mass, the only parameters are the mass and production cross-section of the gluino, and the mass of the LSP. While the Gtt and Gbb models are used both to optimize and interpret the analysis, a further interpretation is provided also in terms of a slightly more complicated model. If we allow also the lightest chargino ($\tilde{\chi}_1^{\pm}$) be kinematically accessible, another decay chain opens, where the virtual stop or sbottom decays to a $\tilde{\chi}_1^{\pm}$: $\tilde{t} \to b\tilde{\chi}_1^-$ and $\tilde{b} \to t\tilde{\chi}_1^-$.

The charge-conjugate processes are also possible, and in both cases the overall decay chain of the gluino leads to $\tilde{g} \to tb\tilde{\chi}_1^-$, which we refer to as Gtb model. The model used to reinterpret this analysis assumes the decay $\tilde{\chi}_1^{\pm} \to \tilde{\chi}_1^0 W^{\pm}$, where the W boson can be off-shell if the mass difference between the $\tilde{\chi}_1^{\pm}$ and the $\tilde{\chi}_1^0$ is not large enough to produce an on-shell W boson. This mass difference becomes an additional parameter of the model, and in this analysis we assume that the $\tilde{\chi}_1^{\pm}$ and the $\tilde{\chi}_1^0$ are almost degenerate (in the simulation, they are generated with a mass difference of 2 GeV); in this case the W boson is virtual and results into soft fermions, which most of the times are not reconstructed. This small mass difference is often verified in models where the $\tilde{\chi}_1^{\pm}$ and the $\tilde{\chi}_1^0$ are part of an approximate $SU(2)$ multiplet, and setting it to a fixed value allows to reduce the number of parameters of the model. This analysis provides an interpretation of the result in terms of the BR of the gluino into the Gtt, Gbb and Gtb models. A schematic diagram of the possible decay chains in this mixed-BR interpretation, in addition to the ones shown in Fig. 8.1, is shown in Fig. 8.2.

These gluino decay chains that lead to final states rich in heavy-flavor quarks are dominant in SUSY models where the squarks from the first two generations are significantly heavier than stop and sbottom, or also in cases when the $\tilde{\chi}_1^0$ is dominated by the higgsino component.

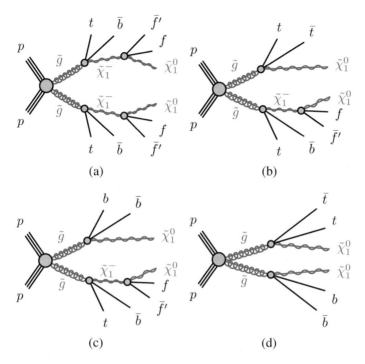

Fig. 8.2 Simplified models used in the reinterpretation of the analysis. **a** Both gluinos have the following decay chain: $\tilde{g} \to tb\tilde{\chi}_1^-$ with $\tilde{\chi}_1^- \to f\bar{f}'\tilde{\chi}_1^0$. **b** One gluino decays as in (**a**) and the other as $\tilde{g} \to t\bar{t}\tilde{\chi}_1^0$. **c** One gluino decays as in (**a**) and the other as $\tilde{g} \to b\bar{b}\tilde{\chi}_1^0$. **c** One gluino decays as $\tilde{g} \to t\bar{t}\tilde{\chi}_1^0$ and the other as $\tilde{g} \to b\bar{b}\tilde{\chi}_1^0$

8.1.1 Signal Cross-Section

The cross-section for the production of a gluino pair at $\sqrt{s} = 13$ TeV, considering all the other SUSY particles decoupled, is shown in Fig. 8.3. This computation is at NLO accuracy in the strong coupling constant, adding the resummation of soft emissions at next-to-leading logarithmic (NLL). To compute the nominal cross-section and its uncertainty, the envelope of the predictions obtained with different choices of factorization scale, renormalization scale and PDF set is built; the nominal cross-section is defined as the midpoint of this envelope, and the uncertainty as half of the envelope width, as described in Ref. [1].

8.2 Discriminating Variables

Most of the kinematic variables used in this analysis to discriminate between signal and background are in common with the higgsino search and have already been

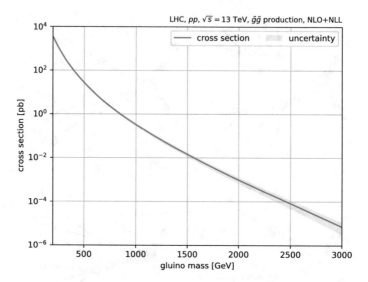

Fig. 8.3 NLO+NLL gluino pair production cross-section with squarks decoupled as a function of mass at $\sqrt{s} = 13$ TeV. Numbers from Ref. [1]

discussed in Sect. 7.6. Two additional variables, used only in this analysis, are defined in this section.

Total jet mass The total jet mass is the sum of the masses of the four leading reclustered jets in the event:

$$M_J^\Sigma = \sum_{i \leq 4} m_{J,i} \; .$$

This variable is particularly useful in the case of boosted Gtt signals, where the four top quarks are produced with high momentum and can be reconstructed as reclustered jets with high mass.

ISR jet In the case of Gbb models, when the mass of the neutralino is close to the mass of the gluino the neutralino is produced almost at rest and the event does not have a sizable E_T^{miss}; this reduces the sensitivity of the analysis to these cases. For these signal models, a sizable value of E_T^{miss} can be present in events where the gluino system is recoiling against a hard jet from ISR; if we assume that the initial-state radiation (ISR) jet is the most energetic jet in the event, there will be an angular separation between this jet and E_T^{miss} larger than the average. The variable $\Delta\phi^{j_1}$, defined as:

$$\Delta\phi^{j_1} = |\Delta\phi(j_1, E_T^{miss})| \; ,$$

helps therefore in selecting signal events where the E_T^{miss} derives from a neutralino boosted by an ISR jet.

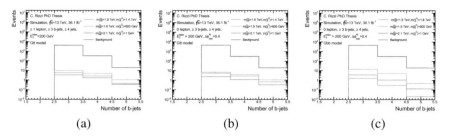

(a) (b) (c)

Fig. 8.4 Distribution of number of *b*-tagged jets in background events and in **a** Gtt signals in a 1-lepton selection, **b** Gtt signals in a 0-lepton selection and **c** Gbb signals in a 0-lepton selection

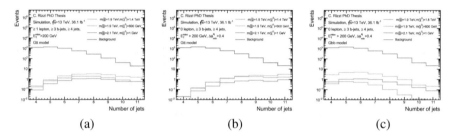

(a) (b) (c)

Fig. 8.5 Distribution of number of jets in background events and in **a** Gtt signals in a 1-lepton selection, **b** Gtt signals in a 0-lepton selection and **c** Gbb signals in a 0-lepton selection

The variables defined above, together with the ones defined in Sect. 7.6, allow to identify a region of the phase space that is enriched in signal events. After selecting events with high E_T^{miss} (>200 GeV), at least four signal jets and at least three *b*-tagged jets, events are divided into the 0-lepton category, which requires a lepton veto and also $\Delta\phi_{min}^{4j} > 0.4$, and the 1-lepton category, with at least one signal lepton. Figures 8.4, 8.5, 8.6, 8.7, 8.8 and 8.9 show the distribution of some important kinematic variables for the sum of the SM backgrounds and for selected signal samples.

From Fig. 8.4 it is possible to see how signal events have in general a higher number of *b*-tagged jets than background events; despite this, this variable alone is not sufficient to reduce the number of background events enough to yield a sizable sensitivity to the signal models of interest.

Figure 8.5 shows the distribution of the number of signal jets. In this case, the result of the comparison between the shape of the distribution in signal and in the main backgrounds depends heavily on the specific signal considered. In particular, Gbb signal models tend to have a lower number of jets than Gtt models; this is expected, since in Gtt models the decay of the top quark to a *b*-quark and a *W* boson can lead to up to 12 jets originating from the hard scattering process, while only four of such jets are present in Gbb signals. Furthermore, Gtt events in the 0-lepton channel have more jets than Gtt-events in the 1-lepton channel, since the leptonic decays of the *W* boson reduce the amount of hadronic activity in the final state. It can be noted that while in SM simulations all-hadronic $t\bar{t}$ events do not enter our

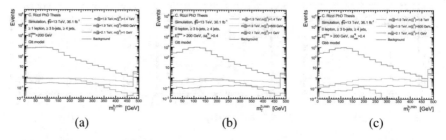

(a)   (b)   (c)

Fig. 8.6 Distribution of $m_{\mathrm{T,min}}^{b\text{-jets}}$ in background events and in **a** Gtt signals in a 1-lepton selection, **b** Gtt signals in a 0-lepton selection and **c** Gbb signals in a 0-lepton selection

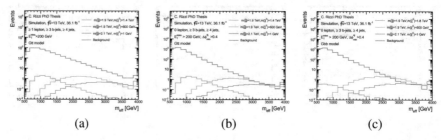

(a)   (b)   (c)

Fig. 8.7 Distribution of m_{eff} in background events and in **a** Gtt signals in a 1-lepton selection, **b** Gtt signals in a 0-lepton selection and **c** Gbb signals in a 0-lepton selection

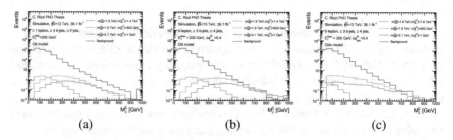

(a)   (b)   (c)

Fig. 8.8 Distribution of M_J^{Σ} in background events and in **a** Gtt signals in a 1-lepton selection, **b** Gtt signals in a 0-lepton selection and **c** Gbb signals in a 0-lepton selection

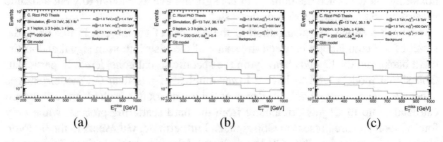

(a)   (b)   (c)

Fig. 8.9 Distribution of $E_{\mathrm{T}}^{\mathrm{miss}}$ in background events and in **a** Gtt signals in a 1-lepton selection, **b** Gtt signals in a 0-lepton selection and **c** Gbb signals in a 0-lepton selection

analysis regions because of the lack of real E_T^{miss}, there are signal events where all the top quarks decay hadronically, since the E_T^{miss} is provided by the neutralino.

A key variable for the suppression of the $t\bar{t}$ background is $m_{T,min}^{b\text{-jets}}$, whose distribution is shown in Fig. 8.6. The kinematic endpoint that characterizes $t\bar{t}$ events is not present for signal events, which tend to have a more uniform distribution. Despite not having a kinematic endpoint, not all signals have an $m_{T,min}^{b\text{-jets}}$ distribution that reaches high values: this is more likely for signals with a large splitting between the gluino and the neutralino masses, which leads to the decay products being boosted.

The m_{eff} distribution for signal and background events is shown in Fig. 8.7. In this case it is possible to observe how for signal models with a low neutralino mass (and therefore a large mass splitting) this variable can be extremely helpful in identifying the presence of signal. In contrast, m_{eff} does not offer a large discriminating power for signals that originate from less energetic final states, as it is the case for the region of the parameter space where the mass of the neutralino approaches the mass of the gluino. A similar argument can be made for M_J^{Σ} and E_T^{miss}, shown in Figs. 8.8 and 8.9 respectively.

8.3 Cut-and-Count Analysis Regions

In this section, we discuss the definition of the cut-and-count analysis regions: first of all the optimization process that leads to the SRs, then the design of the corresponding CRs for the normalization of the $t\bar{t}$ background and of the VRs to test the validity of the background prediction.

8.3.1 Signal Regions

As discussed in Sect. 7.3, the cut-and-count SRs are designed to maximize the significance for specific signal benchmarks, where the significance is defined as the number of Gaussian standard deviations that correspond to the p-value obtained with Eq. 6.13. Since they are not meant to be statistically combined, the SRs do not need to be mutually exclusive and each of them can be optimized independently. The variables considered in the optimization are chosen by selecting the ones that show the most significant differences in shape between signal and background, and are N_{jet}, N_{b-jet}, E_T^{miss}, m_{eff}, M_J^{Σ}, $m_{T,min}^{b\text{-jets}}$ and, in the case of regions requiring at least one lepton, m_T. The SRs are defined as the set of selections that maximize the expected significance for each benchmark model while fulfilling these requirements:

- At least 0.5 expected background events.
- $t\bar{t}$ as major background component, since it is the one that is normalized in CRs and is hence constrained more robustly.
- MC statistical uncertainty in the $t\bar{t}$ component <30%.

• At least 2 expected events for the benchmark signal.

After selecting events that satisfy the E_T^{miss} trigger requirement and have at least four jets, at least three b-jets and $E_T^{miss} > 200$ GeV, three separate optimizations are performed:

Gtt-1L   A first set of regions targets the Gtt signals (shown in Fig. 8.1a) in the final states with at least one reconstructed lepton. Three different SRs are defined in this category: boosted (B), targeting signal models with high gluino mass and a large mass difference between the gluino and the neutralino, medium (M), targeting signal models with intermediate mass splitting, and compressed (C), aiming at signal models where the mass of the neutralino is close to the mass of the gluino.

Gtt-0L   A separate optimization is performed for the cases where the Gtt model leads to final states where there are no reconstructed leptons. Also in this case, three SRs are defined and again referred to as B, M and C depending on the signal models they are targeting. All the Gtt SRs are reported in Table 8.1, together with their control regions (CRs) and SRs whose design is discussed in the next section.

Gbb   The optimization of the SRs targeting the Gbb model (shown in Fig. 8.1b) is performed separately. This signal does not produce any prompt lepton in the final state, but its characteristics differ noticeably from the case of Gtt-0L, particularly due to the lower number of jets. Four SRs are optimized for the Gbb signals: boosted (B), medium (M), compressed (C), and very compressed (VC). The last SR targets compressed Gbb models, where the E_T^{miss} trigger requirement is satisfied because the gluino system is recoiling against an ISR jet, giving additional boost to the neutralinos. To select this topology, this region requires that the leading jet is not b-tagged, and the variable $\Delta\phi^{j_1}$ is included in the optimization. The selections resulting from the optimization of the Gbb SRs are reported in Table 8.2, together with the corresponding CRs and VRs.

8.3.2 Control and Validation Regions

All the CRs of the cut-and-count regions, including the ones for the Gbb SRs and for the Gtt-1L SRs, require the presence of at least one signal lepton. The orthogonality with the Gtt-1L SRs is ensured by applying to all the CRs an upper selection on $m_T < 150$ GeV. To have enough events in the CRs to properly constrain the $t\bar{t}$ background, the $m_{T,min}^{b\text{-jets}}$ selection is removed and other selections are relaxed, so that each CR has a minimum of 10 expected background events.

In the Gtt-1L regions the extrapolation between each pair of CR and SR is validated in two different validation regions (VRs): VR-m_T is designed to verify the extrapolation to high m_T, and is maintained orthogonal to the SR through an inverted M_J^Σ selection. A second region, VR-$m_{T,min}^{b\text{-jets}}$, tests the extrapolation to high $m_{T,min}^{b\text{-jets}}$ by selecting events with high $m_{T,min}^{b\text{-jets}}$ and low m_T; this region is orthogonal to the corre-

Table 8.1 Definitions of the Gtt SRs, CRs and VRs of the cut-and-count analysis. All kinematic variables are expressed in GeV except for $\Delta\phi_{min}^{4j}$, which is in radians. The jet p_T requirement is also applied to b-tagged jets. Table from Ref. [2]

Gtt-1L

Criteria common to all regions: ≥ 1 signal lepton, $p_T^{jet} > 30$ GeV, $N_{b-jet} \geq 3$

Targeted kinematics	Type	N_{jet}	m_T	$m_{T,min}^{b-jets}$	E_T^{miss}	m_{eff}^{incl}	M_J^Σ
Region B (Boosted, Large Δm)	SR	≥ 5	>150	>120	>500	>2200	>200
	CR	$=5$	<150	–	>300	>1700	>150
	VR-m_T	≥ 5	>150	–	>300	>1600	<200
	VR-$m_{T,min}^{b-jets}$	>5	<150	>120	>400	>1400	>200
Region M (Moderate Δm)	SR	≥ 6	>150	>160	>450	>1800	>200
	CR	$=6$	<150	–	>400	>1500	>100
	VR-m_T	≥ 6	>200	–	>250	>1200	<100
	VR-$m_{T,min}^{b-jets}$	>6	<150	>140	>350	>1200	>150
Region C (Compressed, small Δm)	SR	≥ 7	>150	>160	>350	>1000	–
	CR	$=7$	<150	–	>350	>1000	–
	VR-m_T	≥ 7	>150	<160	>300	>1000	–
	VR-$m_{T,min}^{b-jets}$	>7	<150	>160	>300	>1000	–

(continued)

Table 8.1 (continued)

Gtt-0L

Criteria common to all regions: $p_T^{jet} > 30$ GeV

Targeted kinematics	Type	N_{lepton}	N_{b-jet}	N_{jet}	$\Delta\phi^{4j}_{min}$	m_T	$m_{T,min}^{b-jets}$	E_T^{miss}	m_{eff}^{incl}	M_J^Σ
Region B (Boosted, Large Δm)	SR	$= 0$	≥ 3	≥ 7	> 0.4	–	> 60	> 350	> 2600	> 300
	CR	$=1$	≥ 3	≥ 6	–	<150	–	>275	>1800	>300
	VR	$=0$	≥ 3	≥ 6	>0.4	–	–	>250	>2000	<300
Region M (Moderate Δm)	SR	$=0$	≥ 3	≥ 7	>0.4	–	>120	>500	>1800	>200
	CR	$=1$	≥ 3	≥ 6	–	<150	–	>400	>1700	>200
	VR	$=0$	≥ 3	≥ 6	>0.4	–	–	>450	>1400	<200
Region C (Compressed, moderate Δm)	SR	$=0$	≥ 4	≥ 8	>0.4	–	>120	>250	>1000	>100
	CR	$=1$	≥ 4	≥ 7	–	<150	–	>250	>1000	>100
	VR	$=0$	≥ 4	≥ 7	>0.4	–	–	>250	>1000	<100

Table 8.2 Definitions of the Gbb SRs, CRs and VRs of the cut-and-count analysis. All kinematic variables are expressed in GeV except for $\Delta\phi^{4j}_{min}$, which is in radians. The jet p_T requirement is applied to the four leading jets, a subset of which are b-tagged jets. The $j_1 \neq b$ requirement specifies that the leading jet is not b-tagged. Table from Ref. [2]

Gbb

Criteria common to all regions: $N_{jet} \geq 4$, $p_T^{jet} > 30$ GeV

Targeted kinematics	Type	N_{lepton}	N_{b-jet}	$\Delta\phi^{4j}_{min}$	m_T	$m^{b\text{-jets}}_{T,min}$	E_T^{miss}	m_{eff}	Others
Region B (Boosted, Large Δm)	SR	=0	≥3	>0.4	–	–	>400	>2800	–
	CR	=1	≥3	–	<150	–	>400	>2500	–
	VR	=0	≥3	>0.4	–	–	>350	1900–2800	–
Region M (Moderate Δm)	SR	=0	≥4	>0.4	–	>90	>450	>1600	–
	CR	=1	≥4	–	<150	–	>300	>1600	–
	VR	=0	≥4	>0.4	–	>100	250–450	1600–1900	–
Region C (Compressed, small Δm)	SR	=0	≥4	>0.4	–	>155	>450	–	–
	CR	=1	≥4	–	<150	–	>375	–	–
	VR	=0	≥4	>0.4	–	>125	350–450	–	–
Region VC (Very Compressed, very small Δm)	SR	=0	≥3	>0.4	–	>100	>600	–	$p_T^{j1} > 400$, $j_1 \neq b$, $\Delta\phi^{j1} > 2.5$
	CR	=1	≥3	–	<150	–	>600	–	–
	VR	=0	≥3	>0.4	–	>100	225–600	–	

sponding CR thanks to the exclusive jet multiplicity requirement that characterizes the Gtt-1L CRs.

In the case of the Gtt-0L regions, the main extrapolation between each pair of CR and SR is on the number of leptons. This is validated in a specifically-designed 0-lepton VR, that is kept orthogonal to the corresponding SR by an inverted M_J^{Σ} selection.

Similarly, each Gbb region also has a 0-lepton VR, whose orthogonality to the SR is maintained through a shift in the m_{eff} selection for Gbb-B and Gbb-M, and in the E_T^{miss} selection in Gbb-C and Gbb-VC (since these last two regions target

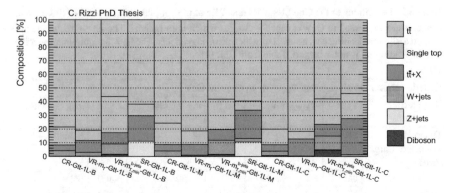

Fig. 8.10 Background composition in the Gtt-1L regions

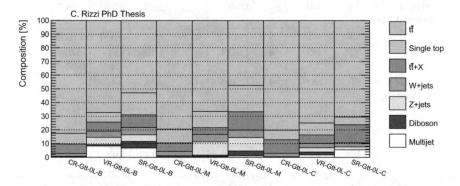

Fig. 8.11 Background composition in the Gtt-0L regions

signal models with a compressed mass spectrum, and therefore do not apply any m_{eff} selection in the SR).

8.3.3 Background Composition

The pre-fit background composition in the cut-and-count analysis regions is shown in Figs. 8.10, 8.11 and 8.12. As can be appreciated, $t\bar{t}$ is the dominant background in all the SRs; the CRs and the VRs have a high $t\bar{t}$ background purity by construction, since they are designed to respectively normalize this background and validate the extrapolation of this normalization in a kinematic regime close to the SRs. The subdominant background contributions are single-top, $t\bar{t}$ +W/Z and, in the 0-lepton channel, $Z(\rightarrow \nu\nu)$+jets.

Figures 8.13, 8.14 and 8.15 show the decay type of the $t\bar{t}$ background for Gtt-1L, Gtt-0L and Gbb regions respectively, classified as described in Sect. 7.7.1. Figure 8.13

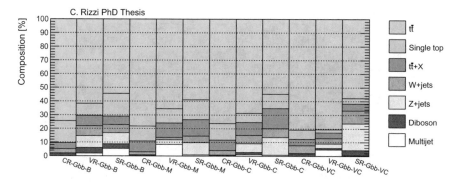

Fig. 8.12 Background composition in the Gbb regions

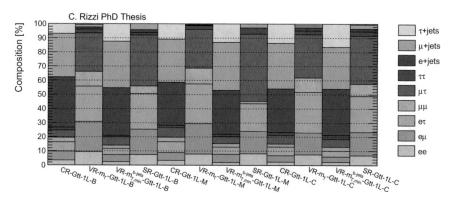

Fig. 8.13 Decay mode of the $t\bar{t}$ background in the Gtt-1L regions

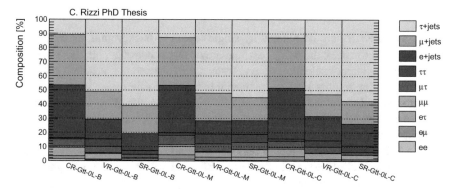

Fig. 8.14 Decay mode of the $t\bar{t}$ background in the Gtt-0L regions

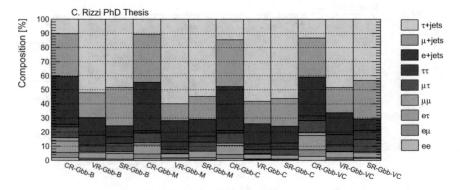

Fig. 8.15 Decay mode of the $t\bar{t}$ background in the Gbb regions

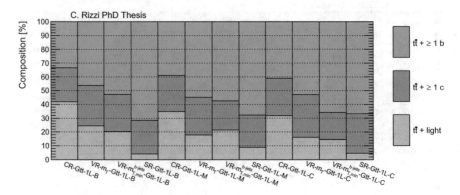

Fig. 8.16 Jet flavor composition in the $t\bar{t}$ background in Gtt-1L regions

shows that the CRs are dominated by single-lepton $t\bar{t}$ background, as well as the VRs designed to check the $m_{\text{T,min}}^{b\text{-jets}}$ extrapolation; this is because both these types of regions have an upper selection on m_{T}. Instead the SRs and the VRs that check the m_{T} extrapolation require high m_{T} values; this selection suppresses single-lepton $t\bar{t}$ background and therefore the SRs and VRs-m_{T} are dominated by dileptonic $t\bar{t}$. In the case of Figs. 8.14 and 8.15 we see that all regions are dominated by single-lepton $t\bar{t}$ background. In these cases, the CRs are 1-lepton regions with an upper cut on m_{T} and the SRs and VRs are 0-lepton regions, so in all of them the main background component is single-lepton $t\bar{t}$. In the case of SRs and VRs though, this single-lepton component is dominated by τ+jets, since this category includes both hadronic and leptonic decays of the τ-lepton.

The classification of the $t\bar{t}$ background based on the flavor of the jets associated to the $t\bar{t}$ production, as described in Sect. 7.7.1, is shown in Figs. 8.16, 8.17 and 8.18. Thanks to the fact that the same selection on the number of b-jets is adopted in the CR, SR and VRs of the same type, the flavor composition is similar in the corresponding

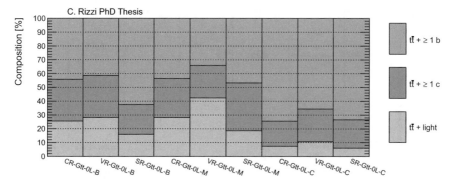

Fig. 8.17 Jet flavor composition in the $t\bar{t}$ background in Gtt-0L regions

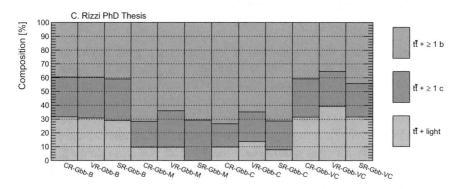

Fig. 8.18 Jet flavor composition in the $t\bar{t}$ background in Gbb regions

regions, even if the fraction of $t\bar{t}$ + HF is sometimes higher in the SRs, especially in the case of Gtt-1L regions.

8.4 Multi-bin Analysis Regions

In this section we describe the definition of the SRs, CRs and VRs of the multi-bin analysis.

8.4.1 Signal Regions

The goal of the multi-bin strategy is to provide a set of regions each optimized for a different signal model but all mutually exclusive, so that they can be statistically combined to increase the sensitivity.

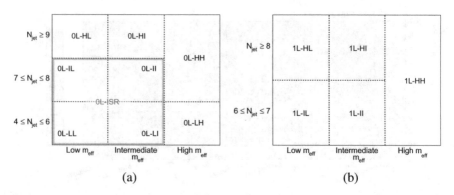

Fig. 8.19 Scheme of the multi-bin analysis for the **a** 0-lepton and **b** 1-lepton regions. The 0L-ISR region is represented with the broad red dashed line in (**a**)

Figure 8.5 shows that the number of jets provides a good handle to separate between Gbb signal models, with lower number of jets, and Gtt signal models, where the number of jets is instead higher; the Gtb model described in Sect. 8.1 will have a number of jets intermediate between the Gtt and the Gtb case, similarly to the mixed-BR case where one of the two produced gluinos decays to $b\bar{b}\tilde{\chi}_1^0$ and the other to $t\bar{t}\tilde{\chi}_1^0$. Within a single decay topology, the variable that best discriminates between signals with different mass splitting between $\tilde{g}$ and $\tilde{\chi}_1^0$ is m_{eff}: as shown in Fig. 8.7, signal models with larger mass splitting tend to have higher values of m_{eff}, since the decay products are more boosted.

These two variables are therefore used, together with the number of signal leptons, to categorize events into mutually exclusive regions. A schematic view of this categorization is given in Fig. 8.19, while the precise numerical values can be found in Tables 8.3, 8.4 and 8.5. The naming convention for the multi-bin SRs is the sequence of number of leptons, category for number of jets and category for m_{eff} regime, where the categories for N_{jet} and m_{eff} are labeled as "L" (low), "I" (intermediate) and "H" (high). So e.g. SR-0L-IL is the SR in the 0-lepton channel with intermediate number of jets and low m_{eff}. Low N_{jet} regions are defined only in the 0-lepton channel, since they target Gbb signal models that do not produce final states with leptons.

A dedicated SR to target Gbb signal models where the gluino pair recoils against an ISR jet is designed following what is done for the cut-and-count analysis. Since the primary target of this ISR region are Gbb models with low mass splitting, its ideal phase space overlaps with that of the 0-lepton regions with low and intermediate N_{jet} and low and intermediate m_{eff}. This SR-ISR relies on the selection of events where the leading jet is very energetic ($p_T > 400$ GeV), is not b-tagged and has a large angular separation with the E_T^{miss} in the event ($\Delta\phi^{j_1} > 2.9$). To allow orthogonality with the ISR regions, all the 0-lepton regions with low and intermediate N_{jet} and low and intermediate m_{eff} are required to have either the leading jet b-tagged or $\Delta\phi^{j_1} < 2.9$.

Table 8.3 Definition of the high-N_{jet} SRs, CRs and VRs of the multi-bin analysis. All kinematic variables are expressed in GeV except for $\Delta\phi_{\min}^{4j}$, which is in radians. Table from Ref. [2]

High-N_{jet} regions

Criteria common to all regions: $N_{b-\text{jet}} \geq 3$, $p_{\text{T}}^{\text{jet}} > 30$ GeV

Targeted kinematics	Type	N_{lepton}	$\Delta\phi_{\min}^{4j}$	m_{T}	N_{jet}	$m_{\text{T,min}}^{b\text{-jets}}$	M_J^Σ	$E_{\text{T}}^{\text{miss}}$	m_{eff}
High-m_{eff} (HH) (Large Δm)	SR-0L	=0	>0.4	–	≥ 7	>100	>200	>400	>2500
	SR-1L	≥ 1	–	>150	≥ 6	>120	>200	>500	>2300
	CR	≥ 1	–	<150	≥ 6	>60	>150	>300	>2100
	VR-0L	=0	>0.4	–	≥ 7	<100 if $E_{\text{T}}^{\text{miss}} > 300$	–	<300 if b-jets $m_{\text{T,min}} > 100$	>2100
	VR-1L	≥ 1	–	>150	≥ 6	<140 if $m_{\text{eff}} > 2300$	–	<500	>2100
Intermediate-m_{eff} (HI) (Intermediate Δm)	SR-0L	=0	>0.4	–	≥ 9	>140	>150	>300	[1800, 2500]
	SR-1L	≥ 1	–	>150	≥ 8	>140	>150	>300	[1800, 2300]
	CR	≥ 1	–	<150	≥ 8	>60	>150	>200	[1700, 2100]
	VR-0L	=0	>0.4	–	≥ 9	<140 if $E_{\text{T}}^{\text{miss}} > 300$	–	<300 if b-jets $m_{\text{T,min}} > 140$	[1650, 2100]
	VR-1L	≥ 1	–	>150	≥ 8	<140 if $E_{\text{T}}^{\text{miss}} > 300$	–	<300 if b-jets $m_{\text{T,min}} > 140$	[1600, 2100]
Low-m_{eff} (HL) (Small Δm)	SR-0L	=0	>0.4	–	≥ 9	>140	–	>300	[900, 1800]
	SR-1L	≥ 1	–	>150	≥ 8	>140	–	>300	[900, 1800]
	CR	≥ 1	–	<150	≥ 8	>130	–	>250	[900, 1700]
	VR-0L	=0	>0.4	–	≥ 9	<140	–	>300	[900, 1650]
	VR-1L	≥ 1	–	>150	≥ 8	<140	–	>225	[900, 1650]

Table 8.4 Definition of the intermediate-N_{jet} SRs, CRs and VRs of the multi-bin analysis. All kinematic variables are expressed in GeV except for $\Delta\phi^{4j}_{\min}$, which is in radians. The $j_1 = b$ requirement specifies that the leading jet is b-tagged. Table from Ref. [2]

Intermediate-N_{jet} regions

Criteria common to all regions: $N_{b-\mathrm{jet}} \geq 3$, $p_T^{\mathrm{jet}} > 30$ GeV

Targeted kinematics	Type	N_{lepton}	$\Delta\phi^{4j}_{\min}$	m_T	N_{jet}	$j_1 = b$ or $\Delta\phi^{ji} \leq 2.9$	$m_{T,\min}^{b\text{-jets}}$	M_J^{Σ}	E_T^{miss}	m_{eff}
Intermediate-m_{eff} (II) (Intermediate Δm)	SR-0L	=0	>0.4	–	[7, 8]	✓	>140	>150	>300	[1600, 2500]
	SR-1L	≥1	–	>150	[6, 7]	–	>140	>150	>300	[1600, 2300]
	CR	≥1	–	<150	[6, 7]	✓	>110	>150	>200	[1600, 2100]
	VR-0L	=0	>0.4	–	[7, 8]	✓	<140	–	>300	[1450, 2000]
	VR-1L	≥1	–	>150	[6, 7]	–	<140	–	>225	[1450, 2000]
Low-m_{eff} (IL) (Low Δm)	SR-0L	=0	>0.4	–	[7, 8]	✓	>140	–	>300	[800, 1600]
	SR-1L	≥1	–	>150	[6, 7]	–	>140	–	>300	[800, 1600]
	CR	≥1	–	<150	[6, 7]	✓	>130	–	>300	[800, 1600]
	VR-0L	=0	>0.4	–	[7, 8]	✓	<140	–	>300	[800, 1450]
	VR-1L	≥1	–	>150	[6, 7]	–	<140	–	>300	[800, 1450]

Table 8.5 Definition of the low-N_{jet} and ISR SRs, CRs and VRs of the multi-bin analysis. All kinematic variables are expressed in GeV except for $\Delta\phi^{4j}_{min}$, which is in radians. The j1 = b (j1 ≠ b) requirement specifies that the leading jet is (not) b-tagged. Table from Ref. [2]

Low-N_{jet} regions

Criteria common to all regions: $N_{b-jet} \geq 3$, $p_T^{jet} > 30$ GeV

Targeted kinematics	Type	N_{lepton}	$\Delta\phi^{4j}_{min}$	m_T	N_{jet}	j1 = b or $\Delta\phi^{j1} \leq 2.9$	p_T^{j4}	$m_{T,min}^{b-jets}$	E_T^{miss}	m_{eff}
High-m_{eff} (LH) (Large Δm)	SR	=0	>0.4	–	[4, 6]	–	>90	–	>300	>2400
	CR	≥1	–	<150	[4, 5]	–	–	–	>200	>2100
	VR	=0	>0.4	–	[4, 6]	–	>90 if E_T^{miss} < 300	–	>200	[2000, 2400]
Intermediate-m_{eff} (LI) (Intermediate Δm)	SR	=0	>0.4	–	[4, 6]	✓	>90	>140	>350	[1400, 2400]
	CR	≥1	–	<150	[4, 5]	✓	>70	–	>300	[1400, 2000]
	VR	=0	>0.4	–	[4, 6]	✓	>90	<140	>300	[1250, 1800]
Low-m_{eff} (LL) (Low Δm)	SR	=0	>0.4	–	[4, 6]	✓	>90	>140	>350	[800, 1400]
	CR	≥1	–	<150	[4, 5]	✓	>70	–	>300	[800, 1400]
	VR	=0	>0.4	–	[4, 6]	✓	>90	<140	>300	[800, 1250]

ISR regions

Criteria common to all regions: $N_{b-jet} \geq 3$, $\Delta\phi^{j1} > 2.9$, $p_T^{j}_1 > 400$ GeV, $p_T^{jet} > 30$ GeV, j1 ≠ b

Type	N_{lepton}	$\Delta\phi^{4j}_{min}$	m_T	N_{jet}	$m_{T,min}^{b-jets}$	E_T^{miss}	m_{eff}
SR	=0	>0.4	–	[4, 8]	>100	>600	<2200
CR	≥1	–	<150	[4, 7]	–	>400	<2000
VR	=0	>0.4	–	[4, 8]	>100	[250, 600]	<2000

In each of these mutually exclusive SR, the selections on variables other than N_{jet}, N_{lepton} and m_{eff} are optimized in a two-steps procedure:

1. First of all, in each region we define the set of selections that maximize the expected significance for a specific benchmark model. These selections have to satisfy the same criteria as of the cut-and-count SRs, described in Sect. 8.3. The benchmark models used to optimize each SR are:

   SR-0L-LL   Gbb, $m(\tilde{g}) = 1.9$ TeV, $m(\tilde{\chi}_1^0) = 1.8$ TeV,
   SR-0L-LI   Gbb, $m(\tilde{g}) = 1.9$ TeV, $m(\tilde{\chi}_1^0) = 600$ GeV,
   SR-0L-LH   Gbb, $m(\tilde{g}) = 2.1$ TeV, $m(\tilde{\chi}_1^0) = 1$ GeV,
   SR-ISR   Gbb, $m(\tilde{g}) = 1.4$ TeV, $m(\tilde{\chi}_1^0) = 1.2$ TeV,
   SR-0L-IL   Gtt, $m(\tilde{g}) = 1.9$ TeV, $m(\tilde{\chi}_1^0) = 1.4$ TeV,
   SR-0L-II   Gtt, $m(\tilde{g}) = 1.9$ TeV, $m(\tilde{\chi}_1^0) = 600$ GeV,
   SR-1L-IL   Gtt, $m(\tilde{g}) = 1.9$ TeV, $m(\tilde{\chi}_1^0) = 1.4$ TeV,
   SR-1L-II   Gtt, $m(\tilde{g}) = 1.9$ TeV, $m(\tilde{\chi}_1^0) = 600$ GeV,
   SR-0L-HL   Gtt, $m(\tilde{g}) = 1.9$ TeV, $m(\tilde{\chi}_1^0) = 1.4$ TeV,
   SR-0L-HI   Gtt, $m(\tilde{g}) = 1.9$ TeV, $m(\tilde{\chi}_1^0) = 600$ GeV,
   SR-0L-HH   Gtt, $m(\tilde{g}) = 2.1$ TeV, $m(\tilde{\chi}_1^0) = 1$ GeV,
   SR-1L-HL   Gtt, $m(\tilde{g}) = 1.9$ TeV, $m(\tilde{\chi}_1^0) = 1.4$ TeV,
   SR-1L-HI   Gtt, $m(\tilde{g}) = 1.9$ TeV, $m(\tilde{\chi}_1^0) = 600$ GeV,
   SR-1L-HH   Gtt, $m(\tilde{g}) = 2.1$ TeV, $m(\tilde{\chi}_1^0) = 1$ GeV.

2. In a second step, we try to make the selections more uniform across all the regions, without penalizing the exclusion power. For each desired simplification of the selection, the exclusion contour is computed and compared to the original one obtained with the selections from the first step. If the simplification of the selections does not bring any significant loss in the exclusion sensitivity, the simplification is adopted. This is the case e.g. for the m_{T} selection, which takes the value of >150 GeV for all the 1-lepton SRs.

8.4.2   Control Regions and Validation Regions

As in the case of the cut-and-count strategy, all the CRs require at least one signal lepton. In the multi-bin strategy the corresponding 0L and 1L SRs share the same CR. This choice is motivated by the observation that 0L and 1L SRs in the multi-bin approach are kinematically closer than Gtt-0L and Gtt-1L regions targeting similar signal topology, and also by the need to have CRs that are either mutually exclusive or completely overlapping to allow the simultaneous inclusion of all the regions in the fit.

Also in the multi-bin approach, the CRs rely on an inverted selection on m_{T} to maintain orthogonality with the corresponding SRs-1L. The requirement on the number of jets is the same as in the SRs-1L, while the SRs-0L require one extra jet.

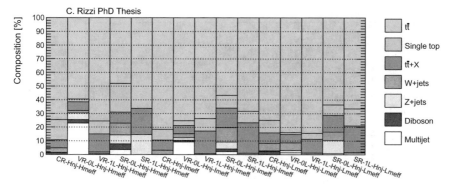

Fig. 8.20 Background composition in the multi-bin regions with high number of jets

Selections on other variables are relaxed to allow enough expected $t\bar{t}$ events in each CR. The extrapolation from the CR to each SR is tested with a dedicated VR. The VRs-0L are orthogonal to the corresponding SRs through inversion of the $m_{\mathrm{T,min}}^{b\text{-jets}}$, m_{eff} or $E_{\mathrm{T}}^{\mathrm{miss}}$ selection; the VRs-1L have the same m_{T} selection as the SRs and are orthogonal to those thanks to an inverted $m_{\mathrm{T,min}}^{b\text{-jets}}$ selection.

To allow enough events in all the VRs, two of them (VR-1L-HI and VR-1L-HL) are not orthogonal. This does not constitute a problem since the VRs are not included in the statistical fit and the overlap has been quantified to be around 15% of the events in these VRs; therefore the validity of the conclusions is not compromised.

8.4.3 Background Composition

The pre-fit background composition in the multi-bin analysis regions is shown in Figs. 8.20, 8.21 and 8.22, for the regions with high, intermediate and low number of jets. Just like in the cut-and-count analysis regions, $t\bar{t}$ is the dominant background in all the SRs, and all the CRs and the VRs have a high $t\bar{t}$ purity.

Figures 8.23, 8.24 and 8.25 show the decay type of the $t\bar{t}$ background for the regions with high, intermediate and low number of jets, classified as described in Sect. 7.7.1. As already noticed when discussing the cut-and-count regions, the SRs-1L and VRs-1L are dominated by dileptonic $t\bar{t}$ since the $m_{\mathrm{T}} > 150\,\mathrm{GeV}$ selection suppresses single-lepton $t\bar{t}$ background, which is instead dominant in the CRs and in the 0-lepton regions. In the case of SRs-0L and VRs-0L the selected $t\bar{t}$ events are mostly the ones where one of the top quarks decays hadronically and the other one to a τ-lepton.

The classification of the $t\bar{t}$ background based on the flavor of the jets associated to the $t\bar{t}$ production, as described in Sect. 7.7.1, is shown in Figs. 8.26, 8.27 and 8.28.

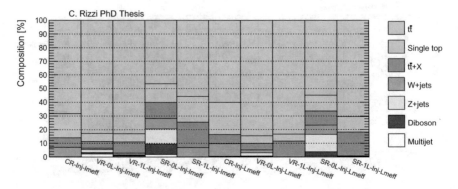

Fig. 8.21 Background composition in the multi-bin regions with intermediate number of jets

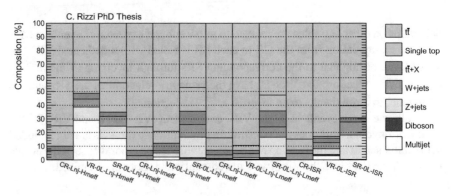

Fig. 8.22 Background composition in the multi-bin regions with low number of jets

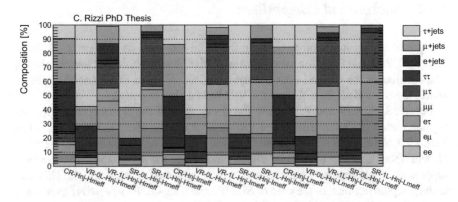

Fig. 8.23 Decay mode of the $t\bar{t}$ background in the multi-bin regions with high number of jets

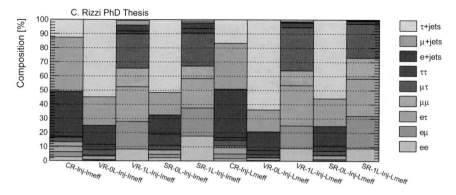

Fig. 8.24 Decay mode of the $t\bar{t}$ background in the multi-bin regions with intermediate number of jets

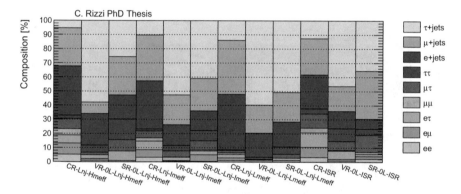

Fig. 8.25 Decay mode of the $t\bar{t}$ background in the multi-bin regions with low number of jets

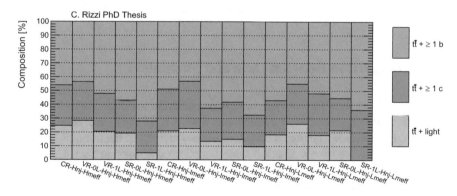

Fig. 8.26 Jet flavor composition in the $t\bar{t}$ background in the multi-bin regions with high number of jets

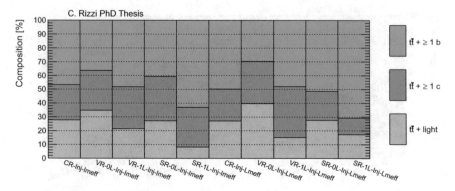

Fig. 8.27 Jet flavor composition in the $t\bar{t}$ background in the multi-bin regions with intermediate number of jets

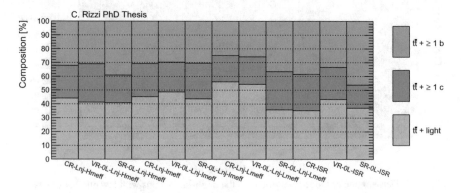

Fig. 8.28 Jet flavor composition in the $t\bar{t}$ background in the multi-bin regions with low number of jets

8.5 Comparison Between Data and Simulation

In this section we show the comparison between data and simulation before the fit in the CR for the same kinematic variables as in Sect. 8.2. All the figures shown in this section do not include systematic uncertainties and include all the relevant MC SFs. The distributions are shown after the same selections discussed in Sect. 8.2, except that the requirement on the number of b-tagged jets is relaxed to $N_{b-\mathrm{jet}} \geq 2$ in order to investigate a region depleted in signal events.

8.5.1 Kinematic Reweighting

The modeling of most kinematic variables related to the energy of the event shows a moderate disagreement when compared with data in the 1-lepton channel, while in the agreement is good in the 0-lepton channel. This is particularly visible in the distribution of m_{eff}, as shown in Fig. 8.29.

While the downward trend in the data/MC ratio in the 1-lepton channel is clearly visible in the bottom panel of Fig. 8.29a, b shows that this is not present in the 0-lepton channel. This difference in trends is problematic for the analysis, since all of the CRs require at least one signal lepton, including the ones used to derive the background prediction for 0-lepton SRs.

To mitigate this problem and to have a better estimate of the background in the high-m_{eff} tail, a reweighting is derived based on the distribution of m_{eff} in a region that requires exactly two b-jets and low $m_{\text{T,min}}^{b\text{-jets}}$ and is therefore orthogonal to all the analysis regions. More details on the reweighting procedure and on its validation are given in Appendix A. All the figures in the rest of this section, as well as the results in the following sections, have been obtained after this reweighting procedure.

8.5.2 Data-MC Comparison After Kinematic Reweighting

Figures 8.30 and 8.31 show the data-MC agreement in the 1-lepton channel after the kinematic reweighting described in the previous paragraph, while the agreement

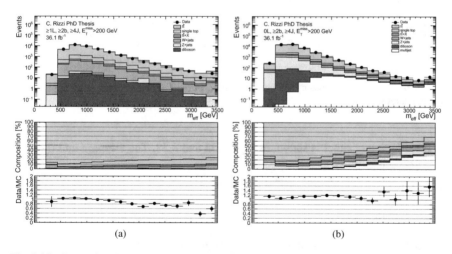

(a)   (b)

Fig. 8.29 Comparison between data and simulation for the distribution of m_{eff} in **a** the 1-lepton channel and **b** the 0-lepton channel

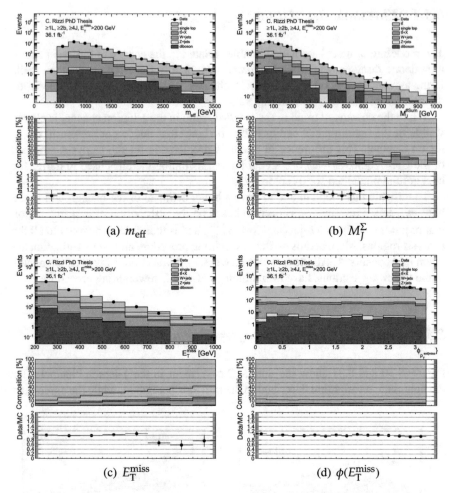

Fig. 8.30 Comparison between data and simulation in the 1-lepton channel, after applying the kinematic reweighting

in the 0-lepton channel is shown in Figs. 8.32 and 8.33. The 1-lepton and 0-lepton preselections require:

- ≥ 4 jets,
- ≥ 2 b-jets,
- $E_T^{miss} > 200$ GeV,

and either at least one signal lepton or a lepton veto and $\Delta\phi_{min}^{4j} > 0.4$.

In Fig. 8.30a it can be noted how the kinematic reweighting improves the data-MC agreement in the 1-lepton channel. All the distributions shown appear well modeled, with the exception of the number of b-jets that shows an underestimate of Monte Carlo (MC) with respect to the data that increases at large b-jet multiplicity, both in

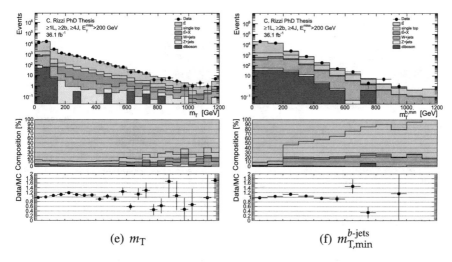

(e) m_T (f) $m_{T,min}^{b\text{-jets}}$

Fig. 8.30 (continued)

the 1-lepton channel and in the 0-lepton channel (Figs. 8.31b and 8.33b respectively). This mis-modelling does not constitute a problem in this analysis, since the CRs have the same b-jet multiplicity as the corresponding SRs.

8.6   Systematic Uncertainties

The sources of systematic uncertainty discussed in Sects. 7.5 and 7.7 are included in the analysis. The relative size of these uncertainties after the fit in the CRs is shown in Fig. 8.34a and b, for the cut-and-count and multi-bin analyses respectively. In the figure, the uncertainties are grouped into:

- Experimental uncertainties, which contain the sum in quadrature of the detector-related uncertainties presented in Sect. 7.5. These are considered for both the background and signal MC samples. The ranking of the different sources of uncertainty changes from region to region; in general the dominant ones are JES (that has a relative impact on the expected background between 4 and 35% in the different SRs), JER (0–26%) and the uncertainties on the b-tagging efficiency and mistagging rate (3–24%).
- Theoretical uncertainties, which are the sum in quadrature of the modeling uncertainties discussed in Sect. 7.7. In this case the dominant uncertainties are the ones on the modeling of the $t\bar{t}$ background, whose impact ranges between 5 and 76%.
- MC statistical uncertainty.
- Statistical uncertainty in the CRs, that is reflected in the uncertainty in the $t\bar{t}$ scale factor and takes values between 10 and 30%.

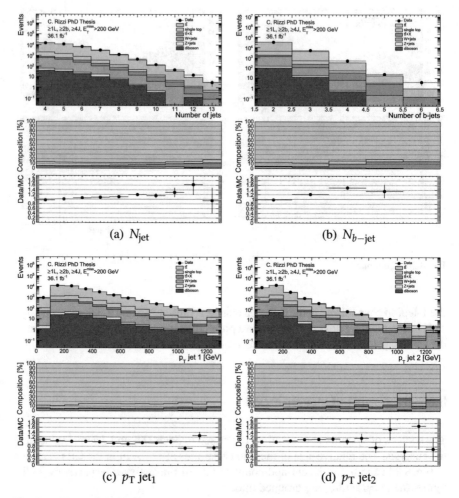

Fig. 8.31 Comparison between data and simulation in the 1-lepton channel, after applying the kinematic reweighting

The total uncertainty (black line in Fig. 8.34) takes into account correlation effects across the different systematic sources, and therefore is not equal to the sum in quadrature of the individual components.

8.7   Results

The statistical procedures discussed in Chap. 6 are used to compare the simulations and the data, and to extract quantitative information on their agreement and on the

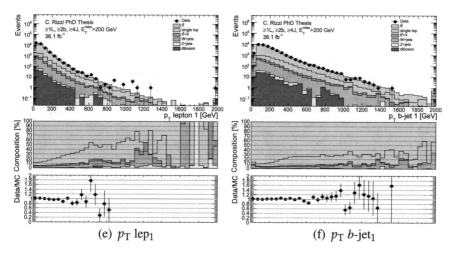

(e) p_T lep$_1$ (f) p_T b-jet$_1$

Fig. 8.31 (continued)

presence of BSM signals. The first step is the background-only fit, a maximum-likelihood fit where only the CRs are included, as described in Sect. 6.4.2. The $t\bar{t}$ normalization is a free parameter of the fit, and is therefore adjusted with the number of observed events in the CRs as constraint. This procedure accounts for potential mis-modellings specific of the kinematic regime close to each SR, which are not necessarily the same for all the regions in the analysis. Therefore each CR is used to derive a normalization factor for the $t\bar{t}$ background, independent from the normalization factors derived in the other CRs, that is then used to derive the background prediction in the corresponding SRs.

The top panel of Fig. 8.35a and b shows the comparison between data and simulation in the CRs of the cut-and-count and multi-bin analyses respectively. In the same figures, the bottom panel shows the $t\bar{t}$ normalization factor derived from the fit in the CRs with the associated uncertainty, driven by the statistical uncertainty in the CRs. The systematic uncertainties on the expected number of events are included in the fit as nuisance parameters. Note that in the case of the cut-and-count analysis the fit is performed separately in each CR, while for the multi-bin analysis all the CRs are included in the same fit, even though with independent parameters for the $t\bar{t}$ normalization.

Figure 8.36a and b show the result of the background-only fit extrapolated to the VRs of the cut-and-count and multi-bin analyses respectively. The upper panel shows the comparison of the number of predicted and observed events, while the bottom panel quantifies the difference between the expected and the observed with the pull, defined as the difference between number of observed and expected events divided by the total uncertainty.

The result of the fit extrapolated to the SRs and the observed number of events in the SRs are finally shown in Fig. 8.37a and b for the cut-and-count and the multi-bin analyses respectively. No significant excess is observed in the SRs; the largest excess

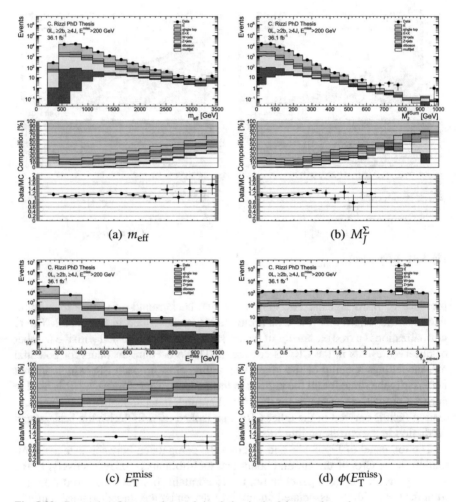

Fig. 8.32 Comparison between data and simulation in the 0-lepton channel

is observed in SR-0L-HH, one of the regions of the multi-bin analysis, where the pull is 2.3 standard deviations. The numerical results in the cut-and-count SRs are summarized in Table 8.6.

8.8 Interpretation

The results described in Sect. 8.7 are used to set limits on the presence of BSM signal models.

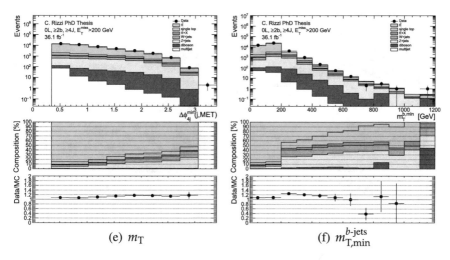

(e) m_T (f) $m_{T,min}^{b\text{-jets}}$

Fig. 8.32 (continued)

8.8.1 Model-Independent Limits

The results of the background-only fit in the cut-and-count SRs are used to place model-independent limits on the number of BSM events in the SRs. This limit is obtained with the CL_s method; for each SR, a fit similar to the background-only fit is performed but the number of observed and expected events in the SRs (with the associated uncertainty) is also included in the fit. Any signal contamination in the CRs is neglected. The limits reported in Table 8.7 are obtained with pseudo-experiments, as discussed in Sect. 6.4.3.

8.8.2 Model-Dependent Limits

The multi-bin analysis is used to place stronger limits on specific signal models. The limit setting procedure is repeated for each signal model, this time considering fully the signal contamination in the CRs and the effect of the modeling and experimental uncertainties on the signal. In this case the results are obtained using the asymptotic approximation [4] when computing the CL_s.

Figure 8.38a and b show the excluded region of the parameter space for Gtt and Gbb models respectively. The dashed blue lines show the expected 95% CL_s limit, which is the limit we would obtain if the observed number of events were identical to the expectation from SM only, and the yellow bands around it indicate the effect of the uncertainty on the background predictions. The observed limits are shown with a solid red line, and the dotted red lines show the impact of the signal systematic uncertainties on the signal modeling (discussed in Sect. 8.1.1). Both in the case

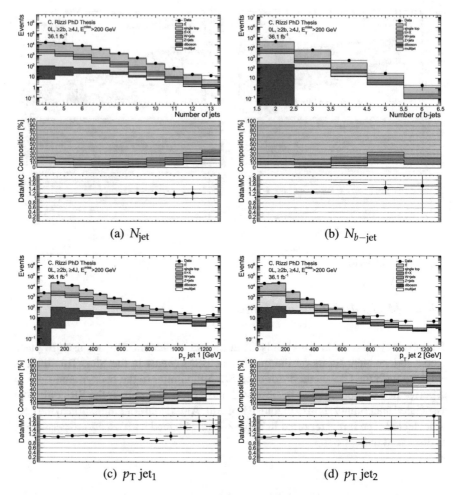

Fig. 8.33 Comparison between data and simulation in the 0-lepton channel

of the Gtt and Gbb models, the observed limit is weaker that the expected. This is particularly evident for Gtt models with massless neutralino, where the limit is driven by SR-0L-HH, which has the largest deviation between expected and observed number of events. The observed exclusion limit for massless neutralino is at around 1.97 and 1.92 for the Gtt and Gbb models respectively; the sensitivity improves by 300 GeV and 450 GeV with respect to the Gbb and Gtt sensitivity obtained from the analysis of the 2015 data [5], shown with dashed black lines in Fig. 8.38. Note that the improvement in observed limits (comparing the solid black and the solid red lines) is much lower than that, since the analysis in Ref. [5] observed a slight deficit while this analysis observes a slight excess.

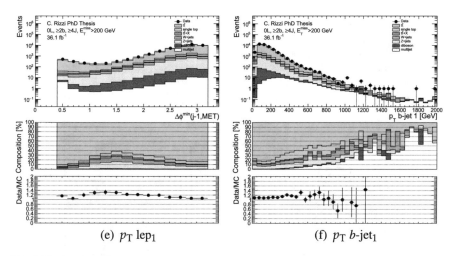

(e) p_T lep$_1$ (f) p_T b-jet$_1$

Fig. 8.33 (continued)

As discussed in Sect. 8.1, the results of the multi-bin analysis are also used to place limits on a more realistic model, where the gluino can decay to $t\bar{t}\tilde{\chi}_1^0$, $b\bar{b}\tilde{\chi}_1^0$, or $t\bar{b}\tilde{\chi}_1^-$, and the $\tilde{\chi}_1^-$ then decays to $\tilde{\chi}_1^0$ and soft fermions. All combinations of different BR to these three decay modes with a unitary constraint on their sum are considered. The inclusion of the gluino BRs as additional parameters makes it impossible to show the results in the two-dimensional plane defined by the masses of the gluino and the neutralino, as it is done for the simpler Gtt and Gbb models. Instead, the limits are presented in the $B(\tilde{g} \to t\bar{t}\tilde{\chi}_1^0) - B(\tilde{g} \to b\bar{b}\tilde{\chi}_1^0)$ plane, assuming that $B(\tilde{g} \to t\bar{b}\tilde{\chi}_1^-) = 1 - B(\tilde{g} \to t\bar{t}\tilde{\chi}_1^0) - B(\tilde{g} \to b\bar{b}\tilde{\chi}_1^0)$, and only a few selected mass points are considered.

Figure 8.39a shows the 95% CL$_s$ exclusion limit for signal models with $m(\tilde{\chi}_1^0) = 1$ GeV and $m(\tilde{g}) = 1.8$, 1.9 and 2.0 TeV. The solid and dashed lines show respectively the observed and expected limits for the different mass hypotheses, distinguished by the different colors. The hashing indicates which side of the plane is excluded. Due to the mild excesses observed in the multi-bin analysis, the observed limits are weaker than the expected. In particular, for a gluino mass of 1.8 TeV, we expect to exclude the entire BR plane, while the "pure Gtb" corner in the bottom left, where both $B(\tilde{g} \to t\bar{t}\tilde{\chi}_1^0)$ and $B(\tilde{g} \to b\bar{b}\tilde{\chi}_1^0)$ are low, is not excluded; none of the points are excluded for a gluino mass of 2.0 TeV.

Figure 8.39b shows instead three signal mass hypotheses with a gluino mass of 1.9 TeV and $m(\tilde{\chi}_1^0) = 1$, 600 and 1000 GeV. Also in this case the observed limits are less stringent than the expected, leaving non-excluded areas also for signals that we expect to exclude for any BR, such as the one with neutralino mass of 600 GeV.

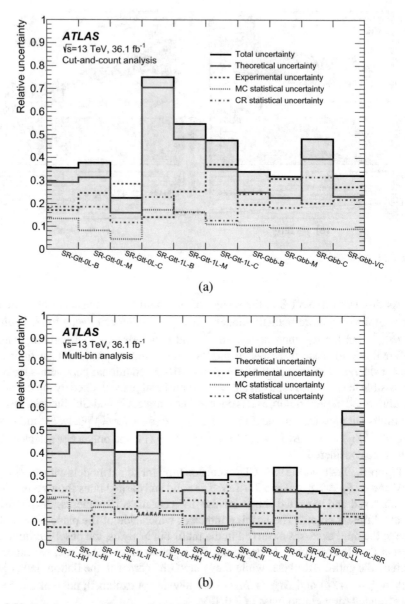

Fig. 8.34 Relative systematic uncertainty in the background estimate for **a** the cut-and-count and **b** the multi-bin analyses. The individual uncertainties can be correlated, such that the total background uncertainty is not necessarily their sum in quadrature. Figure from Ref. [2]

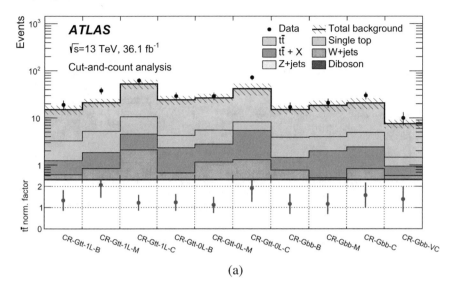

(a)

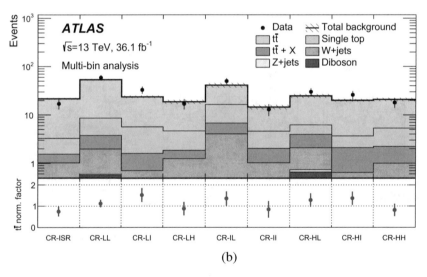

(b)

Fig. 8.35 Pre-fit event yield in CRs and related $t\bar{t}$ normalization factors after the background-only fit for **a** the cut-and-count and **b** the multi-bin analyses. The upper panel shows the observed number of events and the predicted background yield before the fit. The background category $t\bar{t}$ +X includes $t\bar{t}$ +W/Z, $t\bar{t}$ +H and $t\bar{t}t\bar{t}$ events. All of these regions require at least one signal lepton, for which the multijet background is negligible. All uncertainties described in Sect. 8.6 are included in the uncertainty band. The $t\bar{t}$ normalization is obtained from the fit and is displayed in the bottom panel. Figures from Ref. [2]

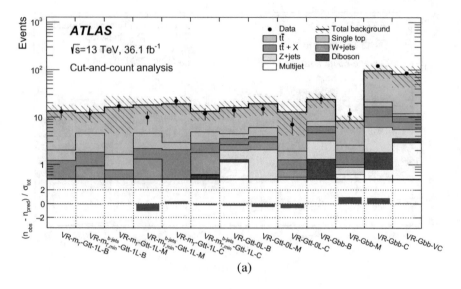

(a)

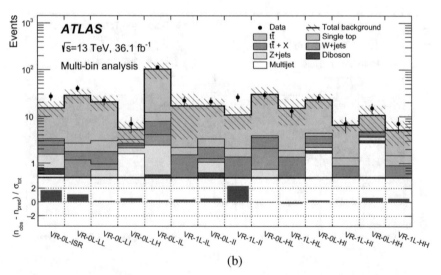

(b)

Fig. 8.36 Results of the background-only fit extrapolated to the VRs of **a** the cut-and-count and **b** the multi-bin analyses. The $t\bar{t}$ normalization is obtained from the fit to the CRs shown in Fig. 8.35. The upper panel shows the observed number of events and the predicted background yield. All uncertainties described in Sect. 8.6 are included in the uncertainty band. The background category $t\bar{t}$ +X includes $t\bar{t}$ +W/Z, $t\bar{t}$ +H and $t\bar{t}t\bar{t}$ events. The lower panel shows the pulls in each VR. Figures from Ref. [2]

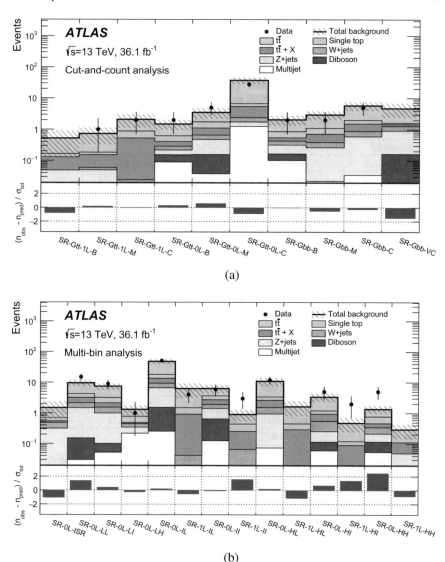

Fig. 8.37 Results of the background-only fit extrapolated to the SRs for **a** the cut-and-count and **b** the multi-bin analyses. The data in the SRs are not included in the fit. The upper panel shows the observed number of events and the predicted background yield. All uncertainties described in Sect. 8.6 are included in the uncertainty band. The background category $t\bar{t} + X$ includes $t\bar{t}W/Z$, $t\bar{t}H$ and $t\bar{t}t\bar{t}$ events. The lower panel shows the pulls in each SR. Figures from Ref. [2]

Table 8.6 Results of the background-only fit extrapolated to the Gtt 1-lepton, Gtt 0-lepton and Gbb SRs in the cut-and-count analysis, for the total background prediction and breakdown of the main background sources. The uncertainties shown include all systematic uncertainties. The data in the SRs are not included in the fit. The background category $t\bar{t} + X$ includes $t\bar{t}W/Z$, $t\bar{t}H$ and $t\bar{t}t\bar{t}$ events. The row "MC-only background" provides the total background prediction when the $t\bar{t}$ normalization is obtained from a theoretical calculation [3]. Table from Ref. [2]

SR-Gtt-1L				
Targeted kinematics	B	M	C	
Observed events	0	1	2	
Fitted background	0.5 ± 0.4	0.7 ± 0.4	2.1 ± 1.0	
$t\bar{t}$	0.4 ± 0.4	0.5 ± 0.4	1.2 ± 0.8	
Single-top	0.04 ± 0.05	0.03 ± 0.06	0.35 ± 0.28	
$t\bar{t} + X$	0.08 ± 0.05	0.09 ± 0.06	0.50 ± 0.28	
Z+jets	0.049 ± 0.023	0.050 ± 0.023	<0.01	
W+jets	<0.01	<0.01	0.024 ± 0.026	
Diboson	<0.01	<0.01	<0.01	
MC-only background	0.43	0.45	1.9	
SR-Gtt-0L				
Targeted kinematics	B	M	C	
Observed events	2	5	28	
Fitted background	1.5 ± 0.5	3.5 ± 1.3	38 ± 8	
$t\bar{t}$	0.9 ± 0.4	1.8 ± 0.7	31 ± 8	
Single-top	0.21 ± 0.14	0.6 ± 0.4	1.3 ± 1.1	
$t\bar{t} + X$	0.12 ± 0.07	0.45 ± 0.25	3.0 ± 1.6	
Z+jets	0.06 ± 0.10	0.3 ± 0.9	0.49 ± 0.31	
W+jets	0.07 ± 0.06	0.18 ± 0.15	0.67 ± 0.22	
Diboson	0.06 ± 0.07	0.12 ± 0.07	<0.01	
Multijet	0.09 ± 0.11	0.04 ± 0.05	1.3 ± 2.1	
MC-only background	1.3	3.3	23	
SR-Gbb				
Targeted kinematics	B	M	C	VC
Observed events	2	2	5	0
Fitted background	2.1 ± 0.7	3.0 ± 1.0	5.8 ± 1.9	4.7 ± 2.3
$t\bar{t}$	1.2 ± 0.6	1.9 ± 0.7	3.8 ± 1.3	3.1 ± 1.3
Single-top	0.31 ± 0.16	0.39 ± 0.16	0.46 ± 0.20	0.15 ± 0.18
$t\bar{t} + X$	0.12 ± 0.06	0.33 ± 0.19	0.6 ± 0.4	0.19 ± 0.11
Z+jets	0.15 ± 0.34	0.2 ± 0.6	0.6 ± 1.3	0.8 ± 1.9
W+jets	0.12 ± 0.09	0.13 ± 0.12	0.29 ± 0.19	0.37 ± 0.30
Diboson	0.06 ± 0.04	<0.01	<0.01	0.15 ± 0.08
Multijet	0.10 ± 0.12	0.022 ± 0.025	0.03 ± 0.04	0.016 ± 0.020
MC-only background	1.9	2.7	4.4	3.9

Table 8.7 The p_0-values and Z (the number of equivalent Gaussian standard deviations), the 95% CL upper limits on the visible cross-section (σ_{vis}^{95}), and the observed and expected 95% CL upper limits on the number of BSM events (S_{obs}^{95} and S_{exp}^{95}). The maximum allowed p_0-value is truncated at 0.5. Table from Ref. [2]

Signal channel	p_0 (Z)	σ_{vis}^{95} [fb]	S_{obs}^{95}	S_{exp}^{95}
SR-Gtt-1L-B	0.50 (0.00)	0.08	3.0	$3.0^{+1.0}_{-0.0}$
SR-Gtt-1L-M	0.34 (0.42)	0.11	3.9	$3.6^{+1.1}_{-0.4}$
SR-Gtt-1L-C	0.50 (0.00)	0.13	4.8	$4.7^{+1.8}_{-0.9}$
SR-Gtt-0L-B	0.32 (0.48)	0.13	4.8	$4.1^{+1.7}_{-0.6}$
SR-Gtt-0L-M	0.25 (0.69)	0.21	7.5	$6.0^{+2.3}_{-1.4}$
SR-Gtt-0L-C	0.50 (0.00)	0.39	14.0	$17.8^{+6.6}_{-4.5}$
SR-Gbb-B	0.50 (0.00)	0.13	4.6	$4.6^{+1.7}_{-1.0}$
SR-Gbb-M	0.50 (0.00)	0.12	4.4	$5.0^{+1.9}_{-1.1}$
SR-Gbb-C	0.50 (0.00)	0.18	6.6	$6.9^{+2.7}_{-1.8}$
SR-Gbb-VC	0.50 (0.00)	0.08	3.0	$4.6^{+2.0}_{-1.3}$

8.9 Comparison of Cut-and-Count and Multi-bin Strategies

In this analysis, the results of the multi-bin analysis are used to set model-dependent limits on selected signal models. Also the regions of the cut-and-count analysis, used to set the model-independent limits in Sect. 8.8.1, can be used to provide a statement on specific signal models. The different regions of the cut-and-count analysis are not orthogonal and therefore cannot be combined in a single statistical fit, but they can still be "visually combined" in a single exclusion contour by selecting for each mass point the cut-and-count region with the best expected sensitivity, and this contour can be compared with the contour obtained with the multi-bin approach. This is shown in Fig. 8.40a and b for the Gtt and Gbb models respectively. Since we want to make a statement on the sensitivity of the two strategies, we compare only the expected exclusion contours (in blue for the cut-and-count analysis, in pink for the multi-bin analysis) and not the observed exclusion contours. The grey numbers on the figures represent the relative difference in expected upper limit on the signal strength: a negative number indicates a stronger limit from the multi-bin analysis. In the case of the Gtt grid in Fig. 8.40a, the multi-bin approach provides an improvement in expected upper limit of ≈30% in the bulk of the mass plane, and up to ≈50% for the more challenging kinematic regime where the mass of the neutralino is closer to the mass of the gluino. In terms of gluino mass reach, this translates in an expected limit about 70 GeV stronger for massless neutralino. In the case of the Gbb signal models, while it is still true that for most of the mass points the expected limit on the signal strength is better with the multi-bin approach, for the region of the parameter space with intermediate gluino and neutralino masses the cut-and-count analysis provides a stronger sensitivity. This is balanced by a better sensitivity of the multi-bin strategy for $m(\tilde{g}) \approx 2$ TeV, $m(\tilde{\chi}_1^0) \approx 700$ GeV (where the cut-and-count approach shows the negative impact of a switch in best-expected region for neighboring mass points), and the two strategies have the same limit for massless neutralino. The reason of the different relative performance of multi-bin and cut-and-count approaches in the Gtt and Gbb signal models relies on the optimization strategy of the multi-bin analysis, which favors Gtt: the regions with intermediate number of jets, which could still be sensitive to Gbb models, are optimized based on Gtt signal benchmarks.

8.10   Update with 2017 Data

The mild excess observed in the multi-bin region SR-0L-HH was verified with the 2017 data as soon as they became available. In particular, the analysis described in the present chapter has been reproduced with all the data collected by the ATLAS experiment in 2015, 2016 and 2017, corresponding to a total integrated luminosity of 79.8 fb^{-1}. Both the cut-and-count and multi-bin regions have been updated in Ref. [6], but in this section we present only the results of the multi-bin regions, since the main purpose is the followup of the excess in the analysis of the 2015–2016 data. Differences in the object definition that occurred due to a change in the ATLAS software release between the two analyses are not discussed in this section and do not impact significantly the physics results; more details can be found in Ref. [6].

8.10.1   Results

The results of the fit in the CRs extrapolated to the VRs and SRs are shown in Figs. 8.41 and 8.42 respectively. We can see that there is good agreement with the predicted background in the VRs, and the SRs do now show any significant deviation from the expectations. In particular, the excess in SR-0L-HH is no longer present.

8.10.2   Interpretation

In this section, we show the model-dependent limits obtained for the update of the multi-bin analysis with 79.8 fb^{-1} of data. Figure 8.43a and b show the exclusion contour at 95% CL for the Gtt and Gbb models respectively; these figures can be compared with the equivalent ones for 36.1 fb^{-1}, Fig. 8.38a and b. The expected limit increases by about 50 GeV and 120 GeV for the Gtt and the Gbb models respectively. The difference is larger for the observed limit: 270 GeV for the Gtt model and 280 GeV for the Gbb model. This is because, while in the 36.1 fb^{-1} analysis the observed limit is weaker than the expected due to a few small excesses in the multi-bin analysis,

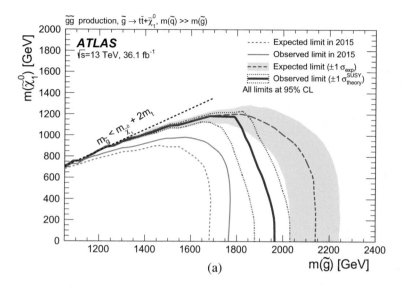

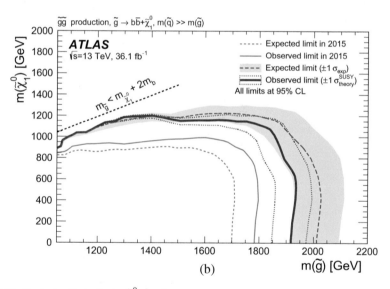

Fig. 8.38 Exclusion limits in the $\tilde{\chi}_1^0$ and $\tilde{g}$ mass plane for the **a** Gtt and **b** Gbb models obtained in the context of the multi-bin analysis. The dashed and solid bold lines show the 95% CL expected and observed limits, respectively. The shaded bands around the expected limits show the impact of the experimental and background uncertainties. The dotted lines show the impact on the observed limit of the variation of the nominal signal cross-section by $\pm 1\sigma$ of its theoretical uncertainty. The 95% CL expected and observed limits from the ATLAS search based on 2015 data [5] are also shown. Figures from Ref. [2]

Fig. 8.39 Exclusion limits in the $\tilde{g} \to t\bar{t}\tilde{\chi}_1^0$ and $\tilde{g} \to b\bar{b}\tilde{\chi}_1^0$ branching ratio plane assuming **a** a neutralino mass of 1 GeV and three gluino masses (1.8, 1.9 and 2.0 TeV) and **b** a gluino mass of 1.9 TeV and three neutralino masses (1, 600 and 1000 GeV). In **a**, the expected limit for a gluino mass of 1.8 TeV follows the plot axes, meaning that the whole plane is expected to be excluded at 95% CL. The dashed and solid bold lines show the 95% CL expected and observed limits, respectively. The hashing indicates which side of the line is excluded. The upper right half of the plane is forbidden by the requirement that the sum of branching ratios does not exceed 100%. Figures from Ref. [2]

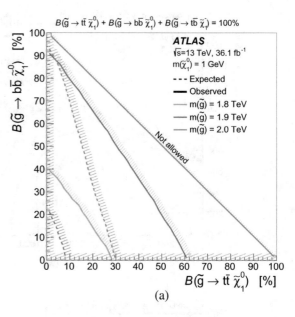

(a)

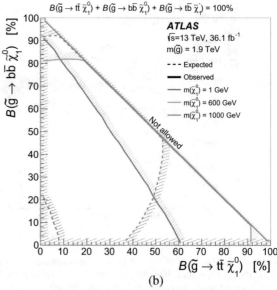

(b)

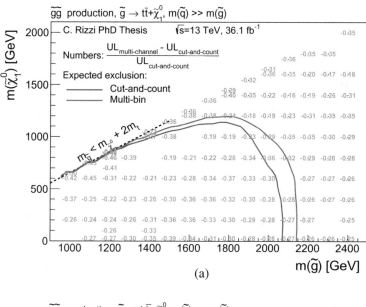

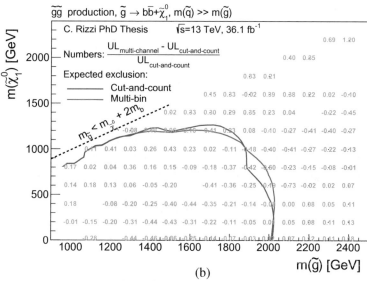

Fig. 8.40 Exclusion limits in the $\tilde{\chi}_1^0$ and $\tilde{g}$ mass plane for the **a** Gtt and **b** Gbb models obtained in the context of the multi-bin analysis (pink line) and of the cut-and-count analysis (blue line). The gray numbers show the relative difference in expected UL on the signal strength between the multi-bin and the cut-and-count analysis: a negative number indicates a lower expected UL for the multi-bin analysis

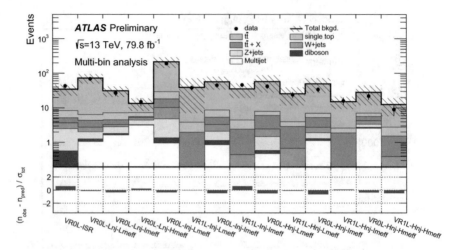

Fig. 8.41 Results of the background-only fit extrapolated to the VRs of the multi-bin analysis. The data in the SRs are not included in the fit. The upper panel shows the observed number of events and the predicted background yield. All the experimental and modeling uncertainties as defined in Ref. [6] are included in the uncertainty band. The background category $t\bar{t} + X$ includes $t\bar{t}W/Z$, $t\bar{t}H$ and $t\bar{t}t\bar{t}$ events. The lower panel shows the pull in each region. Figure from Ref. [6]

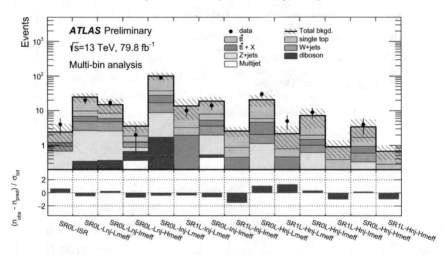

Fig. 8.42 Results of the background-only fit extrapolated to the SRs of the multi-bin analysis. The data in the SRs are not included in the fit. The upper panel shows the observed number of events and the predicted background yield. All the experimental and modeling uncertainties as defined in Ref. [6] are included in the uncertainty band. The background category $t\bar{t} + X$ includes $t\bar{t}W/Z$, $t\bar{t}H$ and $t\bar{t}t\bar{t}$ events. The lower panel shows the pull in each region. Figure from Ref. [6]

in the 79.8 fb^{-1} analysis the observed limits are slightly stronger than the expected in most of the m($\tilde{g}$)-m($\tilde{\chi}_1^0$) plane.

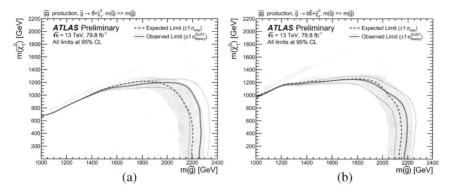

Fig. 8.43 Exclusion limits in the $\tilde{\chi}_1^0$ and $\tilde{g}$ mass plane for the **a** Gtt and **b** Gbb models obtained in the context of the multi-bin analysis. The dashed and solid bold lines show the 95% CL expected and observed limits, respectively. The shaded bands around the expected limits show the impact of the experimental and background uncertainties. The dotted lines show the impact on the observed limit of the variation of the nominal signal cross-section by $\pm 1\sigma$ of its theoretical uncertainty. Figures from Ref. [6]

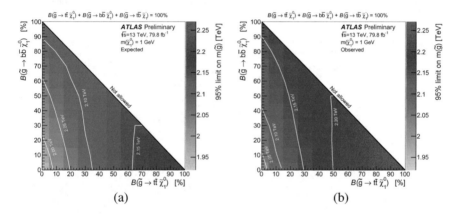

Fig. 8.44 The **a** expected and **b** observed 95% CL exclusion limits on the gluino mass as a function of the gluino branching ratio to Gbb (vertical) and Gtt (horizontal) models. Gluinos not decaying to either the Gtt or Gbb mode are assumed to decay via Gtb instead. In this figure $m_{\tilde{\chi}_1^0}$ is fixed to 1 GeV. The z-axis indicates the maximum excluded gluino mass for each point in the branching ratio space. The white lines indicate contours at mass intervals of 50 GeV. The exclusion limits were derived using the multibin analysis. Figures from Ref. [6]

The interpretation with variable BR of the gluino to $t\bar{t}\tilde{\chi}_1^0$, $b\bar{b}\tilde{\chi}_1^0$, or $t\bar{b}\tilde{\chi}_1^-$ (where the $\tilde{\chi}_1^-$ decays to $\tilde{\chi}_1^0$ and soft fermions) is performed also on the 2015–2017 data but with a different presentation choice: for each point in the BR plane, the highest excluded gluino mass at 95% CL is shown for a specific neutralino mass. Figure 8.44 shows the expected and observed 95% CL exclusion limits for m($\tilde{\chi}_1^0$) = 1 GeV, while Figs. 8.45 and 8.46 assume m($\tilde{\chi}_1^0$) = 600 GeV and 1 TeV respectively. In all the

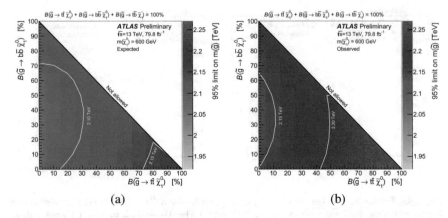

(a)   (b)

Fig. 8.45 The **a** expected and **b** observed 95% CL exclusion limits on the gluino mass as a function of the gluino branching ratio to Gbb (vertical) and Gtt (horizontal) models. Gluinos not decaying to either the Gtt or Gbb mode are assumed to decay via Gtb instead. In this figure $m_{\tilde{\chi}_1^0}$ is fixed to 600 GeV. The z-axis indicates the maximum excluded gluino mass for each point in the branching ratio space. The white lines indicate contours at mass intervals of 50 GeV. The exclusion limits were derived using the multibin analysis. Figures from Ref. [6]

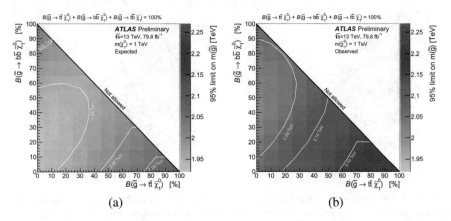

(a)   (b)

Fig. 8.46 The **a** expected and **b** observed 95% CL exclusion limits on the gluino mass as a function of the gluino branching ratio to Gbb (vertical) and Gtt (horizontal) models. Gluinos not decaying to either the Gtt or Gbb mode are assumed to decay via Gtb instead. In this figure $m_{\tilde{\chi}_1^0}$ is fixed to 1000 GeV. The z-axis indicates the maximum excluded gluino mass for each point in the branching ratio space. The white lines indicate contours at mass intervals of 50 GeV. The exclusion limits were derived using the multibin analysis. Figures from Ref. [6]

figures, the white lines indicate contours at mass intervals of 50 GeV. For a given neutralino mass, the strongest limits on the gluino mass are in the pure-Gtt corner and the weakest ones in the pure-Gtb corner for light neutralino mass, while with the increase in neutralino mass the weakest-exclusion point moves to points with higher BR to $b\bar{b}\tilde{\chi}_1^0$.

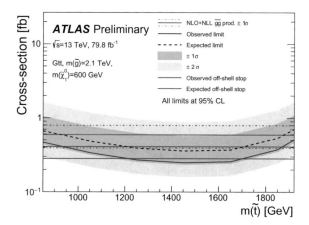

Fig. 8.47 Exclusion limits as a function of $m(\tilde{t})$, in the Gtt model but with an on-shell stop, in the context of the multi-bin analysis. The $\tilde{g}$ and $\tilde{\chi}_1^0$ masses are fixed to 2.1 TeV and 600 GeV respectively. The dashed and solid bold lines show the 95% CL expected and observed limits, respectively. The solid red line indicates the theoretical cross-section for a 2.1 TeV gluino. The thin blue lines indicate the expected (dashed) and observed (solid) limits for the case of the off-shell stop, where $m(\tilde{t}) = 5$ TeV. Figure from Ref. [6]

Two additional interpretations have been included in this version of the analysis, to test the dependence of the limits on the assumption on $m(\tilde{t})$ for the Gtt model and on $m(\tilde{t})$, $m(\tilde{b})$ and m $(\tilde{\chi}_1^\pm)$ for the model with mixed BRs. In the case of the Gtt model, the nominal exclusion limit is provided assuming an off-shell stop: $m(\tilde{t})$ is set to 5 TeV. Figure 8.47 shows the expected and observed 95% cross-section upper limit for a signal model with $m(\tilde{g}) = 2.1$ TeV and $m(\tilde{\chi}_1^0) = 600$ GeV as a function of the stop mass; the thin blue lines show the limit in the case of off-shell stop. The range chosen for $m(\tilde{t})$ allows an on-shell decay for both $\tilde{g} \to \tilde{t}t$ and $\tilde{t} \to t\tilde{\chi}_1^0$. We can see that, while for intermediate values of $m(\tilde{t})$ the sensitivity is similar (and for some masses even better) than the off-shell case, when $m(\tilde{t})$ is close to the kinematic boundary of $m(\tilde{\chi}_1^0)+m(t)$ and $m(\tilde{g})-m(t)$ the sensitivity degrades. This is because when $m(\tilde{t})$ approaches $m(\tilde{\chi}_1^0)+m(t)$, the top quark originating from the $\tilde{t}$ decay has low momentum; instead, when $m(\tilde{t})$ approaches $m(\tilde{g})-m(t)$, it is the top quark produced in the $\tilde{g}$ decay that has less energy than in the off-shell case, limiting the sensitivity of the analysis.

The second additional interpretation is presented in Fig. 8.48, showing the expected and observed 95% CL cross-section upper limit for the mixed-BR model with on-shell stop, sbottom and $\tilde{\chi}_1^\pm$, as a function of the stop mass. The signal model considered has $m(\tilde{g}) = 2.1$ TeV, $m(\tilde{\chi}_1^0) = 600$ GeV, $m(\tilde{b}) = m(\tilde{t})$, and the $\tilde{\chi}_1^\pm$ mass fixed to 1.2 TeV and 180 GeV lower than the stop mass in Fig. 8.48a and b respectively. In the case of fixed $m(\tilde{\chi}_1^\pm)$, the cross-section limit does not display a strong dependence on on $m(\tilde{t})$ and $m(\tilde{b})$ in the range considered for the masses of stop and sbottom. Instead when $m(\tilde{\chi}_1^\pm)$ increases with $m(\tilde{t})$, we can see that the limit weakens at high $m(\tilde{t})$.

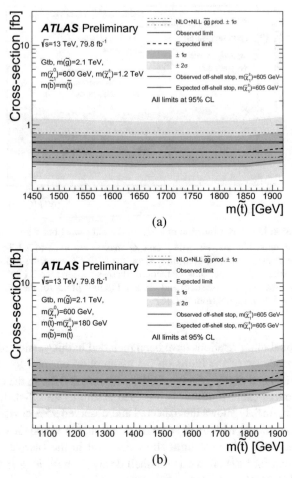

Fig. 8.48 Exclusion limits as a function of $m(\tilde{t})$, in the Gtb model but with an on-shell stop and $\tilde{\chi}_1^\pm$, for **a** for $\tilde{\chi}_1^\pm$ mass fixed to 1.2 TeV, and **b** 180 GeV lower than the stop mass, in the context of the multi-bin analysis. The $\tilde{g}$ and $\tilde{\chi}_1^0$ masses are fixed to 2.1 TeV and 600 GeV respectively. The dashed and solid bold lines show the 95% CL expected and observed limits, respectively. The solid red line indicates the theoretical cross-section for a 2.1 TeV gluino. The thin blue lines indicate the expected (dashed) and observed (solid) limits for the case of the off-shell stop, where $m(\tilde{t}) = 5$ TeV, and $m(\tilde{\chi}_1^\pm) = 605$ GeV. Figures from Ref. [6]

References

1. Borschensky C, Krämer M, Kulesza A, Mangano M, Padhi S, Plehn T, Portell X (2014) Squark and gluino production cross sections in pp collisions at $\sqrt{s}$ = 13, 14, 33 and 100 TeV. Eur Phys J C 74:3174. arXiv:1407.5066 [hep-ph]
2. ATLAS Collaboration (2018) Search for supersymmetry in final states with missing transverse momentum and multiple b-jets in proton-proton collisions at $\sqrt{s}$ = 13 TeV with the ATLAS detector. JHEP 06:107. arXiv:1711.01901 [hep-ex]
3. Czakon M, Mitov A (2014) Top++: a program for the calculation of the top-pair cross-section at hadron colliders. Comput Phys Commun 185:2930. arXiv:1112.5675 [hep-ph]
4. Cowan G, Cranmer K, Gross E, Vitells O (2011) Asymptotic formulae for likelihood-based tests of new physics. Eur Phys J C 71:1554. arXiv:1007.1727 [physics.data-an]. Erratum: Eur Phys J C 73:2501 (2013)
5. ATLAS Collaboration (2016) Search for pair production of gluinos decaying via stop and sbottom in events with b-jets and large missing transverse momentum in pp collisions at $\sqrt{s}$ = 13 TeV with the ATLAS detector. Phys Rev D 94:032003. arXiv:1605.09318 [hep-ex]
6. ATLAS Collaboration (2018) Search for supersymmetry in final states with missing transverse momentum and multiple b-jets in proton-proton collisions at $\sqrt{s}$ = 13 TeV with the ATLAS detector. ATLAS-CONF-2018-041. http://cds.cern.ch/record/2632347

Chapter 9
Search for Higgsino Pair Production

This chapter presents a search for higgsino pair production, with decay to Higgs or Z boson, focusing on the final states with four bottom quarks and E_T^{miss}. Two complementary searches target this signal model: the "high-mass" search, which is the focus of this chapter, is optimized for cases with higgsinos heavy enough to produce sizable E_T^{miss}, while the "low-mass" search targets lower higgsino masses. Section 9.1 describes the signal models used to optimize and interpret the analysis. Section 9.2 discusses the strategy for the reconstruction of the candidate Higgs bosons in the event, and the variables that allow us to separate signal and background events. The definition of the analysis regions are presented in Sects. 9.3 and 9.4, while Sect. 9.5 discusses the background composition. Section 9.6 shows the comparison between data and simulation in a kinematic regime close to that of the analysis regions. Section 9.7 discusses the systematic uncertainties that are included in the analysis, and Sect. 9.8 presents the results. The interpretation of the results in terms of model-independent and model-dependent limits is presented in Sect. 9.9. The low-mass search is briefly described in Sects. 9.10, and 9.11 presents the combined results of the two searches.

9.1 Signal Model

This analysis studies the simplified model shown in Fig. 9.1, targeting primarily the decays via Higgs boson. These are models of GGM [1–5] or GMSB [6, 7] (already discussed in Sect. 2.3.5), where the lightest neutralino is the (NLSP), and it decays promptly to the gravitino ($\tilde{G}$), which is the LSP, and a SM boson.

As discussed in Sect. 2.3.2 SUSY predicts five Higgs bosons. Their superpartners (higgsinos) mix with the superpartners of the electroweak gauge bosons to form charginos and neutralinos.

© Springer Nature Switzerland AG 2020
C. Rizzi, *Searches for Supersymmetric Particles in Final States with Multiple Top and Bottom Quarks with the Atlas Detector*, Springer Theses, https://doi.org/10.1007/978-3-030-52877-5_9

Fig. 9.1 Diagram for the simplified model considered in the analysis. The primary interpretation of the analysis is the decay via Higgs bosons, but decays via varied branching ratios to Z bosons are also studied. The production of the $\tilde{H}$ occurs via mass-degenerate pairs of charginos or neutralinos, which decay to the $\tilde{\chi}_1^0$ and immeasurably low momentum particles

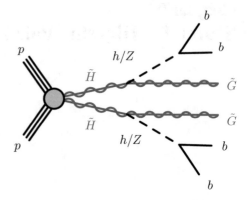

In models where the lightest neutralinos and charginos are dominated by the higgsino component, the four lightest charginos and neutralinos are nearly degenerate [8–10] and ordered as: $m_{\tilde{\chi}_1^0} < m_{\tilde{\chi}_1^\pm} < m_{\tilde{\chi}_2^0}$. These models are particularly interesting because they arise in the limit where $|\mu| < |M_1|, |M_2|$, which is the same limit that minimizes the fine tuning problem in the Higgs sector of the MSSM.

In the case of a higgsino-like neutralino, the direct production of a $\tilde{\chi}_1^0 \tilde{\chi}_1^0$ pair is suppressed, and the production cross-section is dominated by $\tilde{\chi}_1^0 \tilde{\chi}_2^0$, $\tilde{\chi}_1^0 \tilde{\chi}_1^\pm$, $\tilde{\chi}_2^0 \tilde{\chi}_1^\pm$, and $\tilde{\chi}_1^+ \tilde{\chi}_1^-$ production. The $\tilde{\chi}_2^0$ and $\tilde{\chi}_1^\pm$ then decay to the $\tilde{\chi}_1^0$ and soft particles that cannot be detected (originating from the decay of off-shell W and Z bosons), therefore all of these production processes give practically the same final state as $\tilde{\chi}_1^0 \tilde{\chi}_1^0$ pair production.

Since in this chapter we consider only the case where the lightest neutralino is dominated by the higgsino component, we will use interchangeably the notation $\tilde{\chi}_1^0$ or $\tilde{H}$ to indicate it. We consider only the case where the lifetime of the $\tilde{H}$ is very short and it decays promptly to a Higgs boson or a Z boson and the $\tilde{G}$; this is the case when the mass of the mediators of SUSY breaking is relatively small (smaller than $\approx 10^7$ GeV), while for higher mediator masses the NLSP acquires a finite lifetime, and can decay in the detector or pass through the full detector without decaying.

The analysis described in this thesis targets events where the $\tilde{G}$ is produced with enough transverse momentum to lead to sizable E_T^{miss}. This is the case when the higgsinos have intermediate or high mass. This search is not sensitive to events where the higgsino mass is close to the mass of the Higgs boson, and therefore the signal events have very little E_T^{miss} and do not satisfy the E_T^{miss} trigger requirement. These events are the focus of a dedicated analysis which is briefly discussed in Sect. 9.10.

9.1.1 Signal Cross-Section

The signal cross-section is computed for the pure higgsino case, at NLO+NLL precision, assuming that $\tilde{\chi}_1^0$, $\tilde{\chi}_2^0$ and $\tilde{\chi}_1^\pm$ are degenerate and that all the other SUSY particles decouple [11, 12]. The nominal cross-section and its uncertainty are taken from an envelope of two cross-section predictions using different PDF sets, and it is 3830 ± 160 fb at $m(\tilde{H}) = 150$ GeV, while it decreases to 1.8 ± 0.2 fb at $m(\tilde{H}) = 900$ GeV.

9.1.2 Higgsino Decay Modes

This search is optimized to target cases where both $\tilde{H}$ decay promptly to a Higgs boson and a gravitino with 100% BR. This is not the case in realistic models, where the decays of a short-lived higgsino in GGM scenarios can be to a photon, a Z boson or a Higgs boson with BR that depends on the choice of the parameters. As discussed in Ref. [13], the partial width of the three decays of the $\tilde{\chi}_1^0$ is:

$$\Gamma(\tilde{\chi}_1^0 \to \tilde{G} + \gamma) = \frac{1}{2}(s_\beta + \eta c_\beta)^2 \left(\frac{c_W s_W (M_1 - M_2) m_Z}{M_1 M_2} \right)^2 \mathcal{A} \,,$$

$$\Gamma(\tilde{\chi}_1^0 \to \tilde{G} + Z) = \frac{1}{4}(s_\beta + \eta c_\beta)^2 \left(1 - \frac{m_Z^2}{m_{\tilde{\chi}_1^0}^2} \right)^4 \mathcal{A} \,, \qquad (9.1)$$

$$\Gamma(\tilde{\chi}_1^0 \to \tilde{G} + h) = \frac{1}{4}(s_\beta - \eta c_\beta)^2 \left(1 - \frac{m_h^2}{m_{\tilde{\chi}_1^0}^2} \right)^4 \mathcal{A} \,.$$

In Eq. 9.1, $\mathcal{A}$ is a parameter related to the higgsino lifetime:

$$\mathcal{A} = \frac{m_{\tilde{\chi}_1^0}^5}{16\pi F_0^2} \approx \left(\frac{m_{\tilde{\chi}_1^0}}{100 \text{ GeV}} \right)^5 \left(\frac{100 \text{ TeV}}{\sqrt{F_0}} \right)^4 \frac{1}{0.1 \text{ mm}} \,,$$

where s_W and c_W are the sine and cosine of the Weinberg angle, s_β and c_β are the sine and cosine of the angle whose tangent is the ratio of the up-type to down-type Higgs VEV, η represents the relative sign of the coefficients of the two VEVs in the linear combination that constitutes $\tilde{\chi}_1^0$, M_1 and M_2 are the bino and wino mass parameters respectively, and F_0 is the VEV of the SUSY-breaking F-term, the fundamental scale of SUSY breaking. The three BRs for different choices of the parameters are shown in Fig. 9.2.

The decay to photon is relevant only for very low $m(\tilde{H})$, where the other decays are kinematically suppressed. Once all the decays are allowed, the decay is mostly to

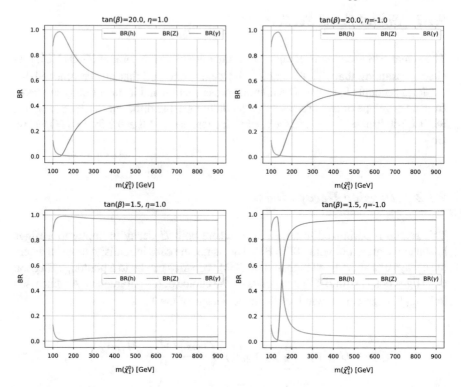

Fig. 9.2 Branching ratios of the higgsino NLSP to a photon, a Z boson, and a Higgs boson, as a function of m_{NLSP}, for $\eta = \pm 1$ and $\tan \beta = 1.5,\ 20$. In all the figures, $M_1 = 500$ GeV, $M_2 = 1000$ GeV, and $m_{h^0} = 125$ GeV. The figures are obtained following the calculations in Ref. [13]

Z boson or Higgs boson; the assumption of $B(\tilde{H} \to h\tilde{G}) = 100\%$, used to optimize this analysis, is realized in the case of relatively low $\tan \beta$ and $\eta = -1$. Instead for low $\tan \beta$ and $\eta = +1$ the dominant decay is to a Z boson, while for high $\tan \beta$ both $B(\tilde{H} \to h\tilde{G})$ and $B(\tilde{H} \to Z\tilde{G})$ tend to converge to the same value for high higgsino mass. This motivates the interpretation of this analysis as a function of $B(\tilde{H} \to h\tilde{G})$, assuming $B(\tilde{H} \to h\tilde{G}) + B(\tilde{H} \to Z\tilde{G}) = 1$, presented in Sect. 9.9.

9.2   Discriminating Variables

A key ingredient of the analysis is the reconstruction of the two Higgs bosons from the decay of the higgsinos. Since this analysis targets events where both Higgs bosons decay to a $b\bar{b}$ pair, the reconstruction starts from four jets, which are chosen according to an ordering that favors b-tagged jets over non-tagged jets, and then orders based on p_T. Practically, this results in the following criteria:

- If there are exactly four b-tagged jets in the event, those are used.
- If there are more than four b-tagged jets, the selected ones are the four b-tagged jets with highest p_T.
- If there are less than four b-tagged jets, the selected ones are the b-tagged jets and the non-tagged jets with highest p_T.

Once the four jets have been selected, they are grouped in two pairs, each one constituting a candidate Higgs bosons. Different algorithms to pair the jets have been tested, and the chosen one is based on minimizing the angular separation between the two jets associated to the same Higgs boson candidate. In particular, the permutation chosen is the one that minimizes:

$$\Delta R^{bb}_{\text{max}} = \max(\Delta R(h_1), \Delta R(h_2)) ,$$

where $\Delta R(h)$ is the distance in the η–ϕ space between the two jets from the same candidate. This choice has a good efficiency in reconstructing the Higgs bosons in the signal, and at the same time avoids creating artificial peaks in the background in correspondence of the Higgs boson mass. More details on the efficiency of the reconstruction and the comparison with other reconstruction algorithms are given in Appendix B.

Beside the variables described in Sect. 7.6, a few other discriminating variables particularly effective for this signal model are described below.

Candidate Higgs bosons The invariant mass of the two Higgs boson candidates built following the procedure outlined above is used as discriminating variable. In particular, we refer to m(h_1) and m(h_2) respectively for the mass of the Higgs candidate with the leading and subleading mass.

The variable $\Delta R^{bb}_{\text{max}}$, defined in the previous section, is used to choose the pairing of the jets while reconstructing the Higgs boson candidates but also as a discriminating variable to separate signal and background.

Modified effective mass In the signal model, only four jets come from the hard scattering process, and are therefore expected to be more energetic than the remaining jets in the event. Therefore the m_{eff} definition is modified to include only the jets that are selected as originating from a Higgs boson; the modified definition is therefore:

$$m^{4j}_{\text{eff}} = \sum_{i=1,...,4} p_T^{j_i} + E_T^{\text{miss}} ,$$

where the sum runs over the jets selected according to the ordering procedure presented above.

The variables defined above, together with the ones defined in Sect. 7.6, allow to identify a region of the phase space that is enriched in signal events. This study is performed after selecting events with high E_T^{miss} (>200 GeV), at least four signal jets, at least three b-tagged jets, zero signal leptons, and $\Delta\phi^{4j}_{\text{min}} > 0.4$.

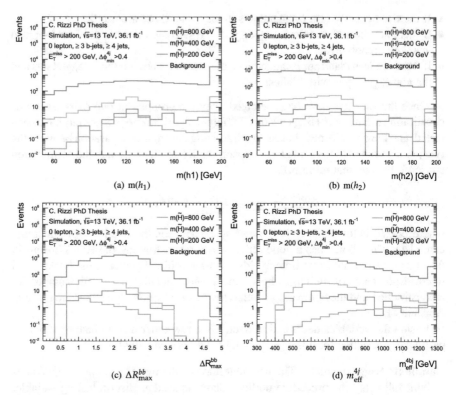

Fig. 9.3 Distribution of the main kinematic variables in background and signal events after the selections described in the text

Figures 9.3 and 9.4 show the distribution of the main kinematic variables for the sum of the SM backgrounds and for different signal samples after these selections. Figure 9.3a and b show the distribution of the mass of the Higgs candidate with the leading and subleading mass respectively. It can be appreciated that the distribution peaks at values around the Higgs mass for signals, while it is flatter for background. The distribution of $\Delta R^{bb}_{\max}$ is shown in Fig. 9.3c. This variable assumes in general a lower value for signal events, in particular for the high-mass signals, where the Higgs bosons are produced with higher p_T and thus have more collimated decay products. Signal events also tend to have a lower number of signal jets, as can be observed in Fig. 9.4a, and they occupy mostly the bins with three and four b-jets in the distribution of $N_{b-\mathrm{jet}}$, shown in Fig. 9.4b. The E_T^{miss} distribution, shown in Fig. 9.4c, displays the expected features: while signals with low-mass higgsinos have low E_T^{miss} values, the distribution tends to assume increasingly high values with the increase of the higgsino mass. A similar feature is shown in Figs. 9.3d and 9.4d for the m_{eff}^{4j} and $m_{\mathrm{T,min}}^{b\text{-jets}}$ distributions respectively: the higher the higgsino mass in the signal, the more signal events differ from background events. While on the one hand this makes it easier to separate them from the SM background, on the other hand the increase in

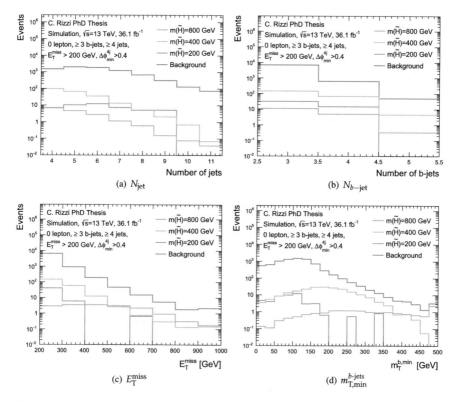

Fig. 9.4 Distribution of the main kinematic variables in background and signal events after the selections described in the text

signal mass implies a decrease in production cross-section, which will be the limiting factor in sensitivity to high-mass signals.

9.3 Signal Regions

This section describes the optimization of the SRs. The high BR of the Higgs boson to a $b\bar{b}$ pair (58%) makes a channel with at least three b-jets very promising to look for signal models in which both higgsinos decay to h+$\tilde{G}$. Therefore, the analysis selections are optimized to maximize the expected sensitivity to signals leading to $hh + E_T^{miss}$ and all the SRs require both boson candidates to have masses compatible with the Higgs mass (the specific mass range is chosen during the optimization).

9.3.1   Multi-bin Regions

The optimization of the multi-bin SRs aims at constructing several orthogonal SRs, that can be statistically combined. The optimization strategy follows these steps:

1. A first variable (var_1) is chosen to define a coarse binning. This variable should provide both a good signal-to-background discrimination and discrimination between signal with different higgsino masses.
2. For each of these bins in var_1, a CR is defined to normalize the $t\bar{t}$ background in a kinematic regime close to the corresponding SR.
3. Each bin based on var_1 is further split based on a second variable, var_2 (in this case all the bins based on the second variable share the same CR).
4. The selections on the remaining kinematic variables are optimized independently in each var_1 bin (i.e. all the regions sharing the same control region (CR) have the same selections on all the variables except var_2).

As described above, var_1 must be able to provide at the same time a good separation between signal and background and a good separation between signals with different $\tilde{H}$ masses. The latter is necessary in order to be able to optimize each bin in var_1 based on a different signal mass, providing in the end a good sensitivity to the entire mass spectrum. In order to understand which variable works best for this scope, the separation algorithm provided by the TMVA toolkit[1] [14] is used, defined as:

$$\text{separation} = \frac{1}{2} \int \frac{(f_s(x) - f_b(x))^2}{f_s(x) + f_b(x)} dx \ , \tag{9.2}$$

where f_s and f_b are the signal and background PDFs of x.

The separation provided by the most promising analysis variables between background and signals with m($\tilde{H}$) = 300, 500 and 800 GeV is shown in Fig. 9.5. It is possible to see that m_{eff}^{4j} is the best variable for our requirements (even if there are individual bins where this is not the case).

Three regular bins with a width of 150 GeV each are defined in m_{eff}^{4j}. The edges of the bins are defined starting from the one with the highest value. As it has already been discussed in Sect. 9.2, signals with high m($\tilde{H}$) have kinematic features that distinguish them clearly from the background, but at the same time a low cross-section (e.g. for m($\tilde{H}$) = 800 GeV, the cross-section is about 3 fb). This makes it more convenient to define the selection that maximize the expected significance for a high-mass benchmark point (800 GeV) as a single-bin region, to concentrate as much as possible the signal events in one single SR, instead of diluting it in several SRs. The optimal selection for the high-m($\tilde{H}$) benchmark signal occupies the highest m_{eff}^{4j} bin, and is determined my maximizing the expected significance over all the other discriminating variables, leading to the selection: $m_{\text{eff}}^{4j} > 1100$ GeV, $m_{\text{T,min}}^{b\text{-jets}} >$

---

[1]TMVA is a toolkit designed for multivariate analyses, this in not the case here: it's only used as a quick way to access the discrimination power of the individual variables.

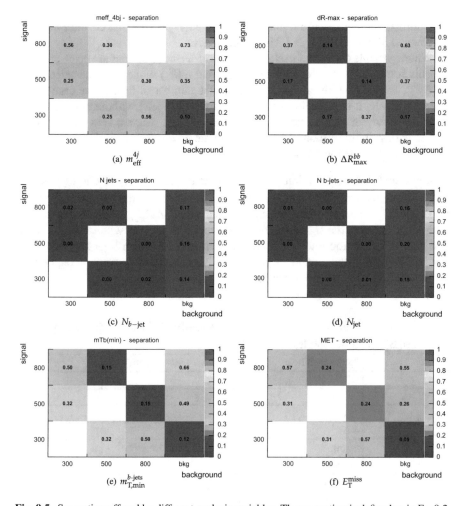

Fig. 9.5 Separation offered by different analysis variables. The separation is defined as in Eq. 9.2

130 GeV, $\Delta R_{\mathrm{max}}^{bb} < 1.4$, $E_{\mathrm{T}}^{\mathrm{miss}} > 200$, ≥ 3 b-jets (77% OP), 4–5 jets, m(h_1) in the range 110–150 GeV, and m(h_2) in the range 90–140 GeV.

Considering the separation power of the different variables shown in Fig. 9.5 and the signal-to-background comparison in Fig. 9.3, $\Delta R_{\mathrm{max}}^{bb}$ is chosen as second binning variable. Once the selections for the high-m_{eff}^{4j} SR have been defined, a similar procedure is repeated for the other m_{eff}^{4j} ranges: 600–850 and 850–1100 GeV. These two bins with lower m_{eff}^{4j} are also split according to the number of b-jets: exactly 3 and ≥ 4. The two bins in b-jet multiplicity are optimized separately. For each of the four remaining regions, the following steps are taken:

1. Optimize for one specific signal benchmark: $m(\tilde{H}) = 300, 500$ GeV for the 600–850 and 850–1100 GeV m_{eff}^{4j} bins respectively.
2. Where possible with a loss in significance $<15\%$, the selections are made uniform. This results to be the case for the $m(h_1)$ and $m(h_2)$ ranges, for the selection in E_T^{miss}, for the first bin in $\Delta R_{\text{max}}^{bb}$, and for the number of jets. In the case of this last variable, it is possible to make it uniform (4–5 jets) without consistent loss in sensitivity in all bins except in the 4b-meff2 bin, where a veto on a 6th jet is penalizing; in this region, the selection on the number of jets is 4–6.
3. Use $\Delta R_{\text{max}}^{bb}$ to define further bins.

It is found that a second bin in $\Delta R_{\text{max}}^{bb}$ is really helpful only in the regions with low and intermediate m_{eff}^{4j}, and ≥ 4 b-jets. This is because the highest m_{eff}^{4j} region is particularly sensitive to high-mass signals, where $\Delta R_{\text{max}}^{bb}$ is small. For signals with lower masses, the signal-to-background ratio in regions with exactly 3 b-jets and high $\Delta R_{\text{max}}^{bb}$ is too low, while the requirement of a fourth b-jet suppresses the SM background further and gives good sensitivity also to the kinematic region with higher $\Delta R_{\text{max}}^{bb}$. The selections that define the SRs are summarized in Table 9.1.

Table 9.1 Signal region definitions for the high-mass analysis. The units of E_T^{miss}, $m_{T,\text{min}}^{b\text{-jets}}$, $m(h_1)$, $m(h_2)$, and m_{eff}^{4j} are GeV. Table from Ref. [15]

	SR-3b-meff1-A	SR-3b-meff2-A	SR-3b-meff3-A	
$N_{b-\text{jet}}$	=3	=3	≥3	
E_T^{miss}	>200			
$\Delta\phi_{\text{min}}^{4j}$	>0.4			
N_{jet}	4–5	4–5	4–5	
$m_{T,\text{min}}^{b\text{-jets}}$	>150	>150	>130	
$m(h_1)$	110–150			
$m(h_2)$	90–140			
$\Delta R_{\text{max}}^{bb}$	0.4–1.4	0.4–1.4	0.4–1.4	
m_{eff}^{4j}	600–850	850–1100	>1100	

	SR-4b-meff1-A	SR-4b-meff1-B	SR-4b-meff2-A	SR-4b-meff2-B
$N_{b-\text{jet}}$	≥4	≥4	≥4	≥4
E_T^{miss}	>200			
$\Delta\phi_{\text{min}}^{4j}$	>0.4			
N_{jet}	4–5	4–5	4–6	4–6
$m_{T,\text{min}}^{b\text{-jets}}$	–	–	–	–
$m(h_1)$	110–150			
$m(h_2)$	90–140			
$\Delta R_{\text{max}}^{bb}$	0.4–1.4	1.4–2.4	0.4–1.4	1.4–2.4
m_{eff}^{4j}	600–850	600–850	850–1100	850–1100

9.3.2 Cut-and-Count Regions

The analysis strategy described in Sect. 9.3.1 leads to a good sensitivity for exclusion across the entire mass spectrum, but the constraint of building orthogonal regions makes it hard to have a high discovery significance for signals with intermediate masses. To solve this problem cut-and-count SRs are defined as well, with the goal of providing robust regions capable of discovering SUSY signatures, and allowing an easier reinterpretation of the results.

The upper selection on m_{eff}^{4j}, that makes the different SRs orthogonal, is the one limiting the most the sensitivity of the individual SRs. In the exclusion fit this is not a problem, as the sensitivity lost in one bin is recovered in the neighbouring one. To define the discovery regions, all the individual SRs are considered without the upper cut on m_{eff}^{4j} and the expected significance for some benchmark signal models is computed in each. The expected significance is computed as discussed in Sect. 6.4.1, assuming 36.1 fb^{-1} of data and a normalization uncertainty of 30% on the background.

Figure 9.6 shows the expected significance for all the multi-bin analysis regions once the upper m_{eff}^{4j} selection is removed. Using only the modified versions of SR-4b-meff1-A, referred to as SR-4b-meff1-A-disc, and SR-3b-meff3 (orange and green line respectively in the figure) allows us to have good expected sensitivity for all the signal considered. This is always larger than three standard deviations for signals with $m(\tilde{H})$ in the 300–600 GeV range, and ≈ 2.5 standard deviations for the signal with $m(\tilde{H}) = 800$ GeV. The definition of SR-4b-meff1-A-disc is reported in Table 9.2.

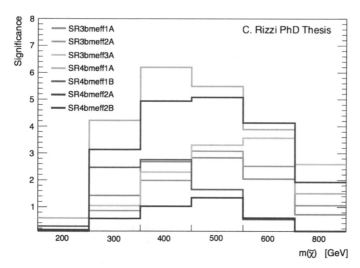

Fig. 9.6 Expected significance for the multi-bin regions after removing the upper m_{eff}^{4j} selection

Table 9.2 Definition of the high-mass analysis SR-4b-meff1-A-disc. The units of E_T^{miss}, $m_{T,min}^{b\text{-jets}}$, $m(h_1)$, $m(h_2)$, and m_{eff}^{4j} are GeV. Table from Ref. [15]

	SR-4b-meff1-A-disc
N_{b-jet}	≥ 4
E_T^{miss}	>200
$\Delta\phi_{min}^{4j}$	>0.4
N_{jet}	4–5
$m_{T,min}^{b\text{-jets}}$	–
$m(h_1)$	110–150
$m(h_2)$	90–140
ΔR_{max}^{bb}	0.4–1.4
m_{eff}^{4j}	>600

9.4   Control and Validation Regions

The $t\bar{t}$ background is normalized in specifically designed CRs. A different CR is built for each bin in m_{eff}^{4j} and b-tagging multiplicity. These CRs are built using side-bands both in $m(h_1)$ and $m(h_2)$. The extrapolation between CRs and SRs is tested in the VRs, which take advantage of events where only one between $m(h_1)$ and $m(h_2)$ is in the SR mass range, as shown in Fig. 9.7.

Some of the selections of the SRs are modified when moving to the CRs or VRs to allow enough events and low signal contamination. The main extrapolation between CRs and SRs are:

Fig. 9.7 The division of signal, control, and validation regions using the $m(h_1)$ and $m(h_2)$ variables in the high-mass analysis

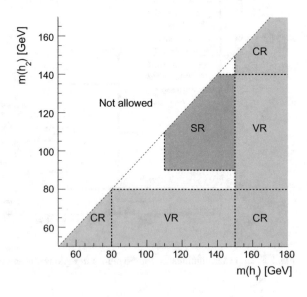

Table 9.3 Control region definitions in the high-mass analysis. The units of E_T^{miss}, $m_{T,\text{min}}^{b\text{-jets}}$, $m(h_1)$, $m(h_2)$, and m_{eff}^{4j} are GeV. Table from Ref. [15]

	CR-3b-meff1	CR-3b-meff2	CR-3b-meff3	CR-4b-meff1	CR-4b-meff2
$N_{b-\text{jet}}$	=3	=3	≥ 3	≥ 4	≥ 4
E_T^{miss}	>200				
$\Delta\phi_{\text{min}}^{4j}$	>0.4				
N_{jet}	4–5	4–5	4–5	4–5	4–6
$m_{T,\text{min}}^{b\text{-jets}}$	>100	>100	>100	–	–
$m(h_1)$, $m(h_2)$	$(m(h_1) <80, m(h_2) <80)$ or $(m(h_1) >150, m(h_2) <80)$ or $(m(h_1) >150, m(h_2) >140)$				
$\Delta R_{\text{max}}^{bb}$	0.4–4	0.4–4	0.4–4	0.4–4	≥ 0.4
m_{eff}^{4j}	600–850	850–1100	>1100	600–850	850–1100

Table 9.4 Validation region definitions in the high-mass analysis. The units of E_T^{miss}, $m_{T,\text{min}}^{b\text{-jets}}$, $m(h_1)$, $m(h_2)$, and m_{eff}^{4j} are GeV. Table from Ref. [15]

	VR-3b-meff1-A	VR-3b-meff2-A	VR-3b-meff3-A	
$N_{b-\text{jet}}$	=3	=3	≥ 3	
E_T^{miss}	>200			
$\Delta\phi_{\text{min}}^{4j}$	>0.4			
N_{jet}	4–5	4–5	4–5	
$m_{T,\text{min}}^{b\text{-jets}}$	>120	>100	>80	
$m(h_1)$, $m(h_2)$	$(80< m(h_1) <150, m(h_2) <80)$ or $(m(h_1) >150, 90< m(h_2) <140)$			
$\Delta R_{\text{max}}^{bb}$	0.4–1.5	0.4–1.7	0.4–1.7	
m_{eff}^{4j}	550–900	800–1150	>1050	
	VR-4b-meff1-A	VR-4b-meff1-B	VR-4b-meff2-A	VR-4b-meff2-B
$N_{b-\text{jet}}$	≥ 4	≥ 4	≥ 4	≥ 4
E_T^{miss}	>200			
$\Delta\phi_{\text{min}}^{4j}$	>0.4			
N_{jet}	4–5	4–5	4–6	4–6
$m_{T,\text{min}}^{b\text{-jets}}$	–			
$m(h_1)$, $m(h_2)$	$(80< m(h_1) <150, m(h_2) <80)$ or $(m(h_1) >150, 90 < m(h_2) <140)$			
$\Delta R_{\text{max}}^{bb}$	0.4–1.7	1.4–3	0.4–1.7	1.4–3
m_{eff}^{4j}	550–900	550–900	800–1150	800–1150

- In the CR, both $m(h_1)$ and $m(h_2)$ are required to be outside the SR mass ranges.
- $\Delta R_{\text{max}}^{bb}$ is relaxed to <4 (or removed).
- $m_{T,\text{min}}^{b\text{-jets}}$ is relaxed by 30 GeV in 3b-meff3 and by 50 GeV in 3b-meff1 and 3b-meff2.

To allow enough events in the validation regions (VRs), the VRs are non orthogonal. Since these regions do not enter the fit, but are only used to validate the background prediction, non-orthogonality is not an issue here. The selections that remove the orthogonality between the different VRs are:

- The edges of m_{eff}^{4j} selection are relaxed by 50 GeV up and down.
- The edges of the $\Delta R_{\mathrm{max}}^{bb}$ bins are relaxed and, when two $\Delta R_{\mathrm{max}}^{bb}$ bins are present for the same type of regions, they are partially overlapping.

With respect to the SRs, the $m_{\mathrm{T,min}}^{b\text{-jets}}$ selection in the VRs (where present) is relaxed as well by 50 GeV. Note that CRs and VRs do not have extrapolation in $N_{b-\mathrm{jet}}$ or N_{jet} with respect to the SRs. The selections for all the CRs and VRs are summarized respectively in Tables 9.3 and 9.4.

9.5 Background Composition

The pre-fit background composition of the analysis regions is shown in Figs. 9.8 and 9.9. It is possible to see how $t\bar{t}$ is the dominant background in all the SRs. The subdominant background contributions are $Z(\to \nu\nu)$+jets and $W(\to \ell\nu)$+jets events. Figures 9.10 and 9.11 show the decay type of the $t\bar{t}$ background, while the flavor composition of the jets that are produced together with the $t\bar{t}$ pair is shown in Figs. 9.12 and 9.13.

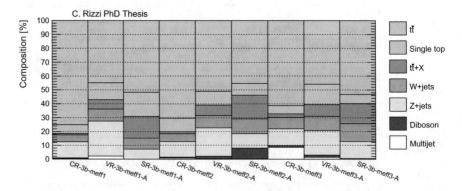

Fig. 9.8 Background composition in the regions with exactly three b-jets

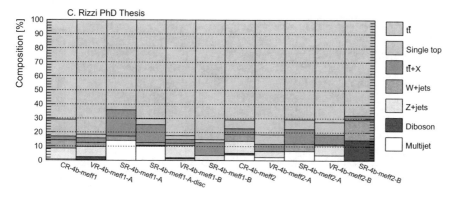

Fig. 9.9 Background composition in the regions with at least four b-jets

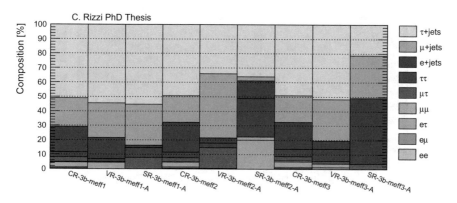

Fig. 9.10 Decay mode of the $t\bar{t}$ background in the regions with exactly three b-jets

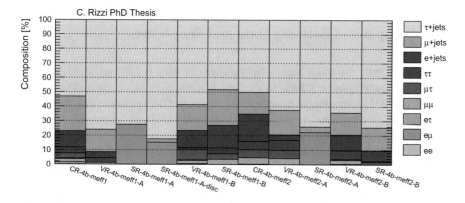

Fig. 9.11 Decay mode of the $t\bar{t}$ background in the regions with at least four b-jets

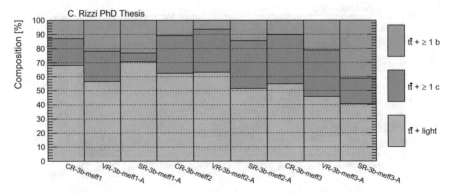

Fig. 9.12 Jet flavor composition of the $t\bar{t}$ background in regions with exactly three b-jets

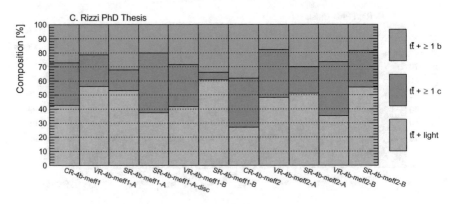

Fig. 9.13 Jet flavor composition of the $t\bar{t}$ background in regions with at least four b-jets

9.6   Comparison Between Data and Simulation

The modeling of the main kinematic variables is very similar to what discussed in Sect. 8.5 for the gluino search, as the few differences in object definitions are not enough to lead to a substantial change in the agreement between data and simulation. This section therefore focuses on the variables specific to Higgs boson reconstruction. The comparison between data and simulation for the variables already shown in Sect. 8.5 but with the object definitions specific to this analysis are shown in Appendix C. There is nevertheless a notable exception: in the analysis described in this chapter, the agreement between data and simulation in the distribution of the number of b-jets is improved, as can be appreciated comparing Fig. 8.33b with Fig. 9.14. This is the result of the improvement in the b-tagging calibration: as discussed in Sect. 5.3.1, the calibration of c-jets used in this analysis is based on $t\bar{t}$ events, rather than on $W + c$ events as in the gluino analysis. The other important difference with respect to the strong-production analysis is that in this case the analysis is performed only

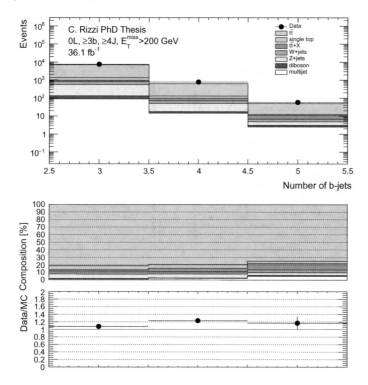

Fig. 9.14 Comparison of the number of b-jets between data and simulation in the preselection described in the text

in regions with a lepton veto; it is not therefore sensitive to the mis-modelling in the 1-lepton channel discussed in Sect. 8.5.1 and no kinematic reweighting is required.

As can be observed in Fig. 9.15, all the variables specific to this analysis shown a good agreement between data and the simulation.

9.7 Systematic Uncertainties

The effect of the systematic uncertainties discussed in Sects. 7.5 and 7.7 is summarized in Fig. 9.16, which shows the relative size of each group of systematic uncertainties after the fit in the CRs. The grouping of the systematic uncertainties is the same as discussed for the gluino search in Sect. 8.6.

For what concerns the experimental uncertainties, the largest ones are those on the JES, whose impact on the expected background yields ranges between 5 and 60% in the different SRs, and on the JER, which impacts the background yields by 10–50%. The uncertainties on the measured b-tagging efficiencies have an impact on the background yields that ranges between 10 and 60%.

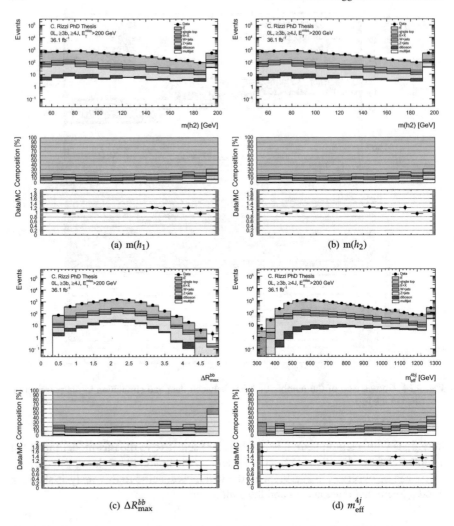

Fig. 9.15 Comparison between data and simulation in the preselection described in the text

The largest theoretical uncertainties are the ones on the modeling of the $t\bar{t}$ background, which range between 10 and 45% for the various regions. Another uncertainty related to the $t\bar{t}$ background is the statistical uncertainty due to the finite size of the CR samples used to derive the $t\bar{t}$ normalization factors, which ranges from 5 to 25% and is represented as a separate contribution in Fig. 9.16. The uncertainties on the modeling of the W/Z+jets backgrounds impact the total yields by 5–20% in the various regions. The modeling uncertainties on the single-top background have a smaller effect on the background yields, leading to changes of at most 11%.

The MC statistical uncertainty reaches 50% in SR-4b-meff1-A, but it is typically lower, around 20%.

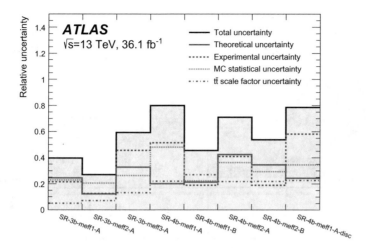

Fig. 9.16 Relative systematic uncertainties on the background estimate for the high-mass analysis. The individual uncertainties can be correlated, such that the total background uncertainty is not necessarily their sum in quadrature. Figure from Ref. [15]

The total uncertainty, which takes into account correlations between the individual uncertainty sources, ranges between 30 and 80% of the total background yields in the different regions.

9.8 Results

Figure 9.17 shows the comparison between data and simulation in the CRs before the fit (top panel) and the scale factor for the $t\bar{t}$ background that is derived from the fit in the CRs (bottom panel). If we compare with the equivalent result for the gluino search, in Fig. 8.35, it is possible to see that on average the $t\bar{t}$ scale factors have values closer to one. This is again because of the improvement in the b-tagging calibration of c-jets was implemented in this analyses. The fit in the CRs is extrapolated to the VRs and to the SRs.

The post-fit data-MC agreement in the VRs is shown in Fig. 9.18: the top panel of this figure shows the post-fit predicted yields in each of the VRs and the data yields, while the bottom panel quantifies the difference between observed data and predictions in terms of the significance, defined as in Ref. [16]. Note that this is different from the pull definition adopted in the gluino search. The closure in the VRs is good: all the bins have discrepancies with significance lower than 0.8.

The results in the SRs are shown in Fig. 9.19. As in Fig. 9.18, the top panel shows the predicted and observed yields in each SR, and the bottom panel the significance of the discrepancy. No significant excess is observed and the observations are in agreement with the SM predictions. The numerical results of the background-only fit

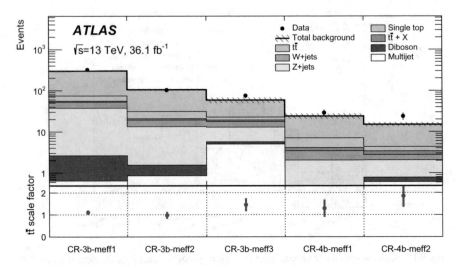

Fig. 9.17 Event yields in control regions and related $t\bar{t}$ normalization factors after the background-only fit for the high-mass analysis. The upper panel shows the observed number of events and the predicted background yield before the fit. All uncertainties shown in Fig. 9.16 are included in the uncertainty band. The background category $t\bar{t} + X$ includes $t\bar{t}W/Z$, $t\bar{t}H$, and $t\bar{t}t\bar{t}$ events. The $t\bar{t}$ normalization is obtained from the fit and is displayed in the bottom panel. Figure from Ref. [15]

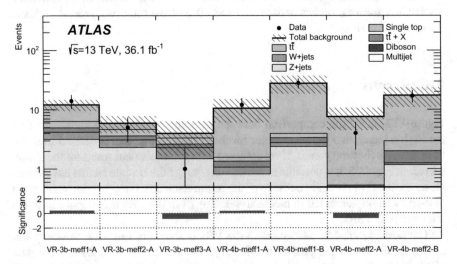

Fig. 9.18 Results of the background-only fit extrapolated to the VRs. The $t\bar{t}$ normalization is obtained from the fit to the CRs shown in Fig. 9.17. The upper panel shows the observed number of events and the predicted background yield. The bottom panel shows the significance of any disagreement between the data and the background model, computed as in Ref. [16]. All uncertainties shown in Fig. 9.16 are included in the uncertainty band. The background category $t\bar{t} + X$ includes $t\bar{t}W/Z$, $t\bar{t}H$, and $t\bar{t}t\bar{t}$ events. Figure from Ref. [15]

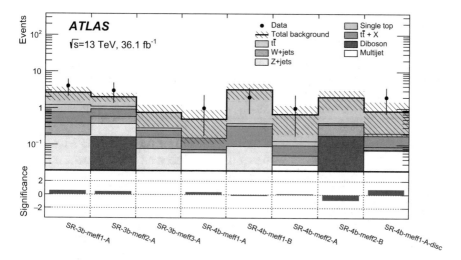

Fig. 9.19 Results of the background only fit extrapolated to the SRs. The $t\bar{t}$ normalization is obtained from the fit to the CRs shown in Fig. 8.35. The data in the SRs are not included in the fit. The upper panel shows the observed number of events and the predicted background yield. The bottom panel shows the significance of any disagreement between the data and the background model, computed as in Ref. [16]. All uncertainties shown in Fig. 9.16 are included in the uncertainty band. The background category $t\bar{t} + X$ includes $t\bar{t}W/Z$, $t\bar{t}H$, and $t\bar{t}t\bar{t}$ events. Figure from Ref. [15]

extrapolated to the SRs are presented in Table 9.5, where the background prediction is also broken down by component. This table shows also the total background prediction before the fit in the CRs, which is labeled as "MC-only background".

9.9 Interpretation

The results presented in Sect. 9.8 are used to set limits on the presence of BSM signals.

9.9.1 Model-Independent Limits

The number of expected and observed events in the two discovery SRs SR-4b-meff1-A-disc and SR-3b-meff3-A are used to set model-independent limits on the number of BSM events. These limits, obtained with the CL_s procedure, ignore any signal contamination in the CRs and are reported in Table 9.6. The same table reports also the model-independent limits obtained from the low-mass analysis, complementary to the analysis discussed in this document, which is briefly presented in Sect. 9.10. To distinguish them from the ones of the low-mass analysis, the results in SR-4b-meff1-

Table 9.5 Results of the background-only fit extrapolated to the SRs of the high-mass analysis, for the total background prediction and breakdown of the main background sources. The uncertainties shown include all systematic uncertainties. The data in the SRs are not included in the fit. The background category $t\bar{t} + X$ includes $t\bar{t}W/Z$, $t\bar{t}H$, and $t\bar{t}t\bar{t}$ events. The row "MC-only background" provides the total background prediction when the $t\bar{t}$ normalization is obtained from a theoretical calculation [17]. Table from Ref. [15]

SR name	SR-3b-meff1-A	SR-3b-meff2-A	SR-3b-meff3-A	SR-4b-meff1-A
N_{obs}	4	3	0	1
Total background	2.6 ± 1.0	2.0 ± 0.5	0.8 ± 0.5	0.5 ± 0.4
Fitted $t\bar{t}$	1.4 ± 0.8	0.89 ± 0.32	0.5 ± 0.4	0.35 ± 0.33
Single top	0.43 ± 0.29	0.17 ± 0.14	0.040 ± 0.017	<0.01
$t\bar{t} + X$	0.39 ± 0.16	0.34 ± 0.14	0.09 ± 0.04	0.08 ± 0.06
Z+jets	0.18 ± 0.14	0.21 ± 0.16	0.07 ± 0.20	<0.01
W+jets	0.20 ± 0.06	0.21 ± 0.09	0.08 ± 0.06	0.013 ± 0.009
Diboson	<0.01	0.16 ± 0.11	<0.01	<0.01
Multijet	<0.01	0.004 ± 0.005	0.004 ± 0.006	0.06 ± 0.05
MC-only background	2.5 ± 1.0	2.0 ± 0.5	0.6 ± 0.4	0.43 ± 0.31
SR name	SR-4b-meff1-B	SR-4b-meff2-A	SR-4b-meff2-B	SR-4b-meff1-A-disc
N_{obs}	2	1	0	2
Total background	3.2 ± 1.5	0.7 ± 0.5	2.0 ± 1.1	0.8 ± 0.7
Fitted $t\bar{t}$	2.8 ± 1.5	0.6 ± 0.5	1.6 ± 1.0	0.6 ± 0.6
Single top	0.06 ± 0.13	0.030 ± 0.019	<0.01	0.030 ± 0.019
$t\bar{t} + X$	0.24 ± 0.10	0.045 ± 0.025	0.039 ± 0.033	0.09 ± 0.06
Z+jets	0.09 ± 0.04	<0.01	<0.01	0.004 ± 0.011
W+jets	<0.01	0.022 ± 0.027	0.18 ± 0.10	0.013 ± 0.008
Diboson	<0.01	<0.01	0.17 ± 0.08	<0.01
Multijet	0.0027 ± 0.0021	0.03 ± 0.04	0.007 ± 0.012	0.07 ± 0.05
MC-only background	2.6 ± 0.9	0.43 ± 0.27	1.3 ± 0.6	0.7 ± 0.5

A-disc and SR-3b-meff3-A are labeled as high-SR-4b-meff1-A-disc and high-SR-3b-meff3-A respectively.

9.9.2   Model-Dependent Limits

A combined fit that includes simultaneously all the CRs and all the orthogonal SRs (i.e. all the SRs except for SR-4b-meff1-A-disc) is used to place limits on the specific models described in Sect. 9.1.

Table 9.6 For each discovery region, the number of observed events (N_{obs}), the number of predicted events (N_{pred}), and 95% CL upper limits on the visible cross-section (σ_{vis}^{95}) and on the number of signal events (S_{obs}^{95}) are shown. The fifth column (S_{exp}^{95}) shows the 95% CL upper limit on the number of signal events given the expected number (and $\pm 1\sigma$ excursions of the expectation) of background events. The last column indicates the discovery p-value ($p(s = 0)$) in significance units. The p-values are capped at 0.5. Results are obtained with 20 000 pseudoexperiments. Table from Ref. [15]

Signal channel	N_{obs}	N_{pred}	σ_{vis}^{95} [fb]	S_{obs}^{95}	S_{exp}^{95}	p_0 (Z)
high-SR-4b-meff1-A-disc	2	0.8 ± 0.7	0.15	5.5	$4.2^{+1.3}_{-0.4}$	0.15 (1.02)
high-SR-3b-meff3-A	0	0.8 ± 0.5	0.08	3.0	$3.1^{+1.2}_{-0.1}$	0.50 (0.00)
low-SR-MET0-meff440	1063	1100 ± 25	2.3	56	79^{+31}_{-23}	0.50 (0.00)
low-SR-MET150-meff440	17	12 ± 8	0.90	22	19^{+5}_{-4}	0.21 (0.80)

The signal model that is used to optimize the analysis regions is higgsino pair production with BR($\tilde{H} \rightarrow h\tilde{G}$) = 100%. The 95% CL upper limit on the total pair production cross-section for this model is shown in Fig. 9.20, as a function of $m(\tilde{H})$. The expected exclusion is between 250 and 830 GeV in $m(\tilde{H})$. Due to the slight deficit in the region with the highest m_{eff}^{4j} selection, the observed exclusion is up to 880 GeV.

A second interpretation of the results is provided in Fig. 9.21, which shows the exclusion contour in the plane BR($\tilde{H} \rightarrow h\tilde{G}$)–$m(\tilde{H})$, with the assumption BR($\tilde{H} \rightarrow h\tilde{G}$)+BR($\tilde{H} \rightarrow Z\tilde{G}$) = 1. For $m(\tilde{H})$ = 400 GeV, which is the mass point with the lowest excluded σ/σ_{theory}, we exclude at 95% CL BRs as low as 45%.

9.10 Complementary Low-Mass Higgsino Search

The analysis discussed in this chapter is limited in sensitivity for low $m(\tilde{H})$, as it is clear from Fig. 9.20. This is because when $m(\tilde{H})$ approaches the Higgs mass, the decay products (Higgs boson and $\tilde{G}$) have increasingly low p_T, and a low-p_T $\tilde{G}$ does not produce enough E_T^{miss} to satisfy the E_T^{miss} trigger requirements. To gain sensitivity also to the low-$m(\tilde{H})$ part of the mass spectrum, which is particularly interesting for naturalness arguments (see the discussion in Sect. 2.3.3), this analysis is complemented by a second analysis that targets low-E_T^{miss} events, referred to as "low-mass" analysis [15]. Events are selected using b-jet triggers and are required

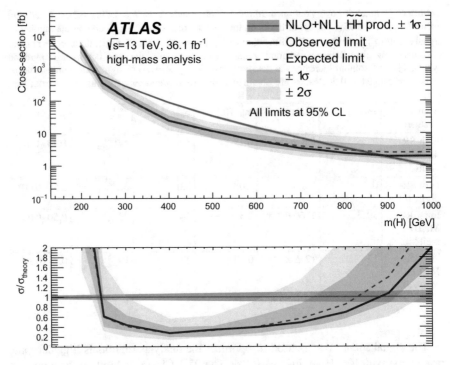

Fig. 9.20 The observed (solid black) versus expected (dashed black) 95% upper limits on the total pair production cross-section for degenerate higgsinos as a function of $m(\tilde{H})$. The 1σ and 2σ uncertainty bands are shown as green and yellow, respectively. The theory cross-section is shown in the red curve. The bottom panel shows the ratio of the observed and expected limits with the theory cross-section. Figure from Ref. [15]

to have at least four b-tagged jets, using a b-tagging OP with an efficiency of 70% (tighter than the 77% used in the high-mass analysis). This analysis uses data from 2016 where the b-jet triggers are available, corresponding to an integrated luminosity of 24.3 fb^{-1}.

The jets used to reconstruct the Higgs candidates are the four with the highest b-tagging score, and are paired minimizing the quantity D_{hh}, defined as:

$$D_{hh} = \left| m_{2j}^{\text{lead}} - \frac{120}{110} m_{2j}^{\text{subl}} \right|,$$

where m_{2j}^{lead} and m_{2j}^{subl} are the masses of the Higgs boson candidates with leading and subleading p_T respectively. This pairing choice tends to create two Higgs candidates with similar mass; for low higgsino masses the b-jets originating form the decay of the Higgs bosons are less collimated, and this choice is therefore more effective than minimizing $\Delta R_{\text{max}}^{bb}$.

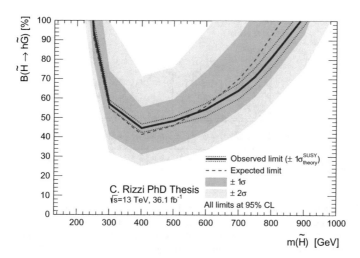

Fig. 9.21 The observed (solid) versus expected (dashed) 95% limits in the $m(\tilde{H})$–BR($\tilde{H} \rightarrow h\tilde{G}$) plane, where BR($\tilde{H} \rightarrow h\tilde{G}$) denotes the branching ratio for the decay $\tilde{H} \rightarrow h\tilde{G}$. The 1σ uncertainty band is overlaid in green and the 2σ in yellow. The regions above the lines are excluded by the analysis

The main background is constituted by multijet events and a small fraction of $t\bar{t}$ events, as opposed to the high mass analysis, where multijet is an almost-negligible background after applying the $\Delta\phi^{4j}_{min}$ selection. The background from $t\bar{t}$ events is further reduced by requiring $X_{Wt} > 1.8$, where X_{Wt} is defined as:

$$X_{Wt} = \sqrt{\left(\frac{m_W - 80.4 \text{ Ge V}}{0.1 m_W}\right)^2 + \left(\frac{m_t - 172.5 \text{ Ge V}}{0.1 m_t}\right)^2}.$$

In the above equation, the top and W-boson candidates are built as described in Ref. [15]. A low value of X_{Wt} corresponds to a high probability of the event to be a $t\bar{t}$ event.

The SR is defined by requiring:

$$X^{SR}_{hh} = \sqrt{\left(\frac{m^{lead}_{2j} - 120 \text{ Ge V}}{0.1 m^{lead}_{2j}}\right)^2 + \left(\frac{m^{subl}_{2j} - 110 \text{ Ge V}}{0.1 m^{subl}_{2j}}\right)^2} < 1.6 ,$$

where $0.1 m^{lead}_{2j}$ and $0.1 m^{subl}_{2j}$ approximate the mass resolution of the two Higgs boson candidates.

The events in the SR are further binned based on the two-dimensional distribution of E^{miss}_T and m_{eff}, and this is used as input in the statistical analysis. The binning used is:

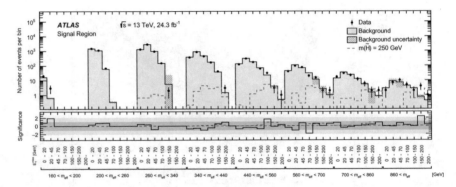

Fig. 9.22 The unrolled distribution of E_T^{miss} and m_{eff}^{4j} for data, background and an example signal sample in the signal region of the low-mass analysis. The bottom panel shows the significance of any disagreement between the data and the background model [16]. The dashed line includes the signal contribution and defines the significance as signal/σ. Figure from Ref. [15]

$$E_T^{miss} = \{0, 20, 45, 70, 100, 150, 200\} \,,$$
$$m_{eff}^{4j} = \{160, 200, 260, 340, 440, 560, 700, 860\} \,,$$

where the values are expressed in GeV.

Two dedicated discovery regions have optimized selections to maximize the discovery significance for $m(\tilde{H}) = 150$ GeV and $m(\tilde{H}) = 300$ GeV:

- low-SR-MET0-meff440: $m_{eff}^{4j} > 440$ GeV.
- low-SR-MET150-meff440: $m_{eff}^{4j} > 440$ GeV, $E_T^{miss} > 150$ GeV.

The background estimate is fully data-driven and relies on a sample with exactly two b-tagged jets (orthogonal to the SR and with very low signal contamination). Selections on m_{2j}^{lead} and m_{2j}^{subl} are used to define a CR and two VRs, both in the $\geq 4b$ and in the $2b$ samples; all these regions exclude the $X_{hh}^{SR} < 1.6$ area, to be orthogonal to the SR. The 2-tag and 4-tag CRs are used to derive a reweighting function to go from the 2-tag sample to the 4-tag sample, that consists in two steps: first of all an overall normalization correction is applied, and then a reweighting based on boosted decision trees corrects for further kinematic differences. This reweighting procedure is tested in the VRs and then applied to the VRs. More details on the background estimate and its validation are available in Ref. [15].

The number of expected and observed events in the SRs of the low-mass analysis is shown in Fig. 9.22, whose bottom panel shows the significance of any disagreement between the data and the background model. A mild excesses is present in the bin with $860 < m_{eff}^{4j} < 2000$ GeV and $150 < E_T^{miss} < 200$ GeV, where four events are observed while the predicted background is 1.0 ± 0.2 events.

9.11 Combined Results

Figures 9.23 and 9.24 show the combined results of the two analyses for the model-dependent exclusion, respectively in the case where $BR(\tilde{H} \rightarrow h\tilde{G}) = 100\%$ and in the $m(\tilde{H})$–$BR(\tilde{H} \rightarrow h\tilde{G})$ plane. The results of the low-mass analysis are used below 300 GeV, while above it is the high-mass search that provides the nominal result. The transition at 300 GeV is chosen such that in the transition point the two analyses have similar sensitivity in the case where $BR(\tilde{H} \rightarrow h\tilde{G}) = 100\%$. In the low-mass analysis the high-E_T^{miss} bins of the SR show a mild excess; therefore, the observed limit is weaker than expected and the portion of the mass spectrum between 230 and 290 GeV is not excluded for $BR(\tilde{H} \rightarrow h\tilde{G}) = 100\%$, despite the expected sensitivity.

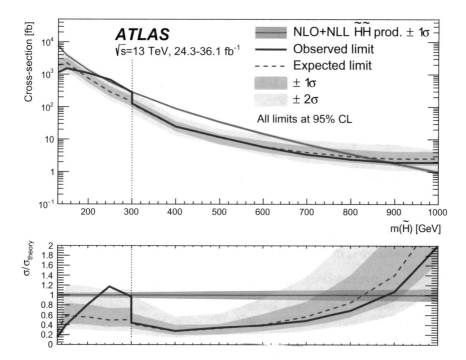

Fig. 9.23 The observed (solid) versus expected (dashed) 95% upper limits on the $\tilde{H}$ pair production cross-section as a function of $m(\tilde{H})$. The 1σ and 2σ uncertainty bands on the expected limit are shown as green and yellow, respectively. The theory cross-section and its uncertainty are shown in the solid and shaded red curve. The results of the low-mass analysis are used below $m(\tilde{H}) = 300$ GeV, while those of the high-mass analysis are used above. Figure from Ref. [15]

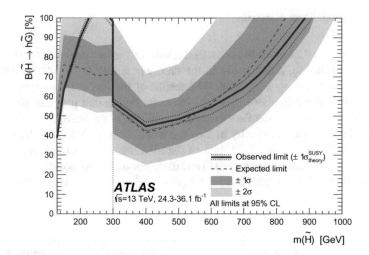

Fig. 9.24 The observed (solid) versus expected (dashed) 95% limits in the $m(\tilde{H})$–$\mathrm{BR}(\tilde{H} \rightarrow h\tilde{G})$ plane, where $\mathrm{BR}(\tilde{H} \rightarrow h\tilde{G})$ denotes the branching ratio for the decay $\tilde{H} \rightarrow h\tilde{G}$. The 1σ uncertainty band is overlaid in green and the 2σ in yellow. The results of the low-mass analysis are used below $m(\tilde{H}) = 300$ GeV, while those of the high-mass analysis are used above. The regions above the lines are excluded by the analyses. Figure from Ref. [15]

References

1. Meade P, Seiberg N, Shih D (2009) General gauge mediation. Prog Theor Phys Suppl 177:143. arXiv:0801.3278 [hep-ph]
2. Cheung C, Fitzpatrick AL, Shih D (2008) (Extra)ordinary gauge mediation. JHEP 07:054. arXiv:0710.3585 [hep-ph]
3. Dine M, Fischler W (1982) A phenomenological model of particle physics based on supersymmetry. Phys Lett B 110:227
4. Alvarez-Gaume L, Claudson M, Wise MB (1982) Low-energy supersymmetry. Nucl Phys B 207:96
5. Nappi CR, Ovrut BA (1982) Supersymmetric extension of the SU(3) x SU(2) x U(1) model. Phys Lett B 113:175
6. Dimopoulos S, Dine M, Raby S, Thomas SD (1996) Experimental signatures of low energy gauge-mediated supersymmetry breaking. Phys Rev Lett 76:3494. arXiv:hep-ph/9601367 [hep-ph]
7. Matchev KT, Thomas SD (2000) Higgs and Z boson signatures of supersymmetry. Phys Rev D 62:077702. arXiv:hep-ph/9908482 [hep-ph]
8. Papucci M, Ruderman JT, Weiler A (2012) Natural SUSY endures. JHEP 09:035. arXiv:1110.6926 [hep-ph]
9. Barbieri R, Pappadopulo D (2009) S-particles at their naturalness limits. JHEP 10:061. arXiv:0906.4546 [hep-ph]
10. Han Z, Kribs GD, Martin A, Menon A (2014) Hunting quasidegenerate Higgsinos. Phys Rev D 89:075007. arXiv:1401.1235 [hep-ph]
11. Fuks B, Klasen M, Lamprea DR, Rothering M (2012) Gaugino production in proton-proton collisions at a center-of-mass energy of 8 TeV. JHEP 10:081. arXiv:1207.2159 [hep-ph]
12. Fuks B, Klasen M, Lamprea DR, Rothering M (2013) Precision predictions for electroweak superpartner production at hadron colliders with Resummino. Eur Phys J C 73:2480. arXiv:1304.0790 [hep-ph]

13. Meade P, Reece M, Shih D (2010) Prompt decays of general neutralino NLSPs at the Tevatron. JHEP 05:105. arXiv:0911.4130 [hep-ph]
14. Hocker A et al. TMVA - toolkit for multivariate data analysis. arXiv:physics/0703039 [physics.data-an]
15. ATLAS Collaboration (2018) Search for pair production of higgsinos in final states with at least three b-tagged jets in $\sqrt{s} = 13$ TeV pp collisions using the ATLAS detector, Submitted to: Phys Rev. arXiv:1806.04030 [hep-ex]
16. Choudalakis G, Casadei D (2012) Plotting the differences between data and expectation. Eur Phys J Plus 127:25. arXiv:1111.2062 [physics.data-an]
17. Czakon M, Mitov A (2014) Top++: a program for the calculation of the top-pair cross-section at hadron colliders. Comput Phys Commun 185:2930. arXiv:1112.5675 [hep-ph]

Chapter 10
Comparison with Other ATLAS and CMS Searches

In this chapter we present the results of the two searches discussed in Chaps. 8 and 9 in the context of the wider program of SUSY searches carried out by the ATLAS and CMS collaborations.

10.1 Gluino Pair Production with Decay Through Third-Generation Squarks

In the ATLAS Collaboration only the gluino search presented in this thesis (referred to as "multi-b search" in the following) provides an interpretation for the Gbb model discussed in Chap. 8. Instead, in the case of the Gtt model, other analyses are interpreted to provide sensitivity to this signal grid. The search described in Ref. [1] targets this model by selecting final states with two same-charge leptons (electrons or muons), high jet multiplicity (at least six jets), one or two b-tagged jets and different E_T^{miss} selections (ranging from no E_T^{miss} selection to $E_T^{miss} > 200\,\text{GeV}$).

Figure 10.1 shows the overlay of the limits obtained with the multi-b search and with the two-same-charge-leptons search. The latter has a very good sensitivity when the neutralino mass approaches the gluino mass, limiting the amount of E_T^{miss} in the final state. This analysis considers also signal models where the mass difference between the gluino and the neutralino is not enough to produce two on-shell top quarks and one of them is off-shell; this is why the limit extends above the "diagonal" where $m(\tilde{g}) = m(\tilde{\chi}_1^0) + 2m(t)$. Also the search targeting final states with large jet multiplicity and E_T^{miss}, described in Ref. [3], is interpreted to provide limits on the Gtt model, that result to be weaker than the ones set by the multi-b and the two-same-charge-leptons searches.

© Springer Nature Switzerland AG 2020
C. Rizzi, *Searches for Supersymmetric Particles in Final States
with Multiple Top and Bottom Quarks with the Atlas Detector*, Springer Theses,
https://doi.org/10.1007/978-3-030-52877-5_10

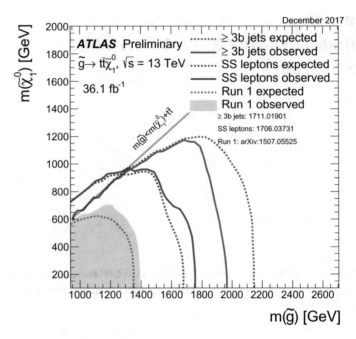

Fig. 10.1 Exclusion limits at 95% CL based on 13 TeV data in the (gluino, lightest neutralino) mass plane for the Gtt simplified model where a pair of gluinos decays promptly via off-shell top squarks to four top quarks and two lightest neutralinos. Theoretical signal cross-section uncertainties are not included in the limits shown. "≥3 jets" refers to the multi-*b* search. Figure from Ref. [2]

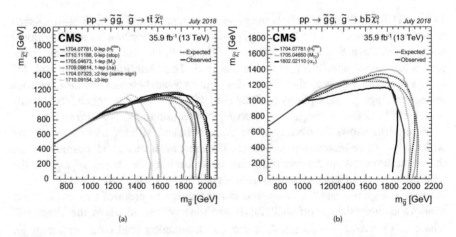

Fig. 10.2 Mass limits at 95% CL obtained for simplified models of gluino pair production with gluino decays to **a** pairs of top quarks and the LSP and **b** pairs of bottom quarks and the LSP. In the figures, the solid (dashed) lines correspond to the observed (median expected) limits. The arXiv numbers corresponding to the different analyses are shown in the legend. Figures from Ref. [9]

In the case of the CMS Collaboration, several analyses provide an interpretation for both the Gtt and the Gbb signal models, shown respectively in Fig. 10.2a and b. The H_T^{miss} analysis [4] provides the best sensitivity for both models in the case of massless neutralino; this analysis performs a four-dimensional scan of 0-lepton events with at least two jets, binning them based on the scalar sum of the p_T of the signal jets in the event (H_T), the negative vector sum of the jet transverse momenta (H_T^{miss}), the number of jets and the number of b-jets.

In the case of the Gtt model, for high gluino mass and intermediate neutralino mass the most sensitive analysis is the 0-lepton analysis with reconstructed hadronically decaying top quarks [5], that builds several SRs based on the number of jets, the number of b-jets, the number of reconstructed top quark candidates, E_T^{miss}, H_T, and m_{T2} (a transverse-mass-type variable designed to reduce the $t\bar{t}$ background). For signal models with small mass difference between the gluino and the neutralino, the most sensitive analysis is also for CMS the same-sign analysis [6], which selects events with two leptons with the same charge and builds several SRs based on number of jets, number of b-jets, E_T^{miss}, H_T, and m_T.

The m_{T2} analysis [7], that uses 0-lepton events with high m_{T2}, is the one most sensitive for Gbb models with high gluino mass and intermediate neutralino mass. The Gbb signal does not produce any prompt lepton, so in this case also when the neutralino mass approaches the kinematic limit it is a 0-lepton analysis that provides the best sensitivity: in the α_T analysis [8], all the SRs have a selection on α_T, defined as the ratio between the energy of the subleading jet and the transverse mass between the leading and subleading jet.

If we compare the ATLAS and CMS results for this signal model, they overall provide similar sensitivity. For massless neutralino, in the case of the Gtt signal the ATLAS multi-b analysis has an expected sensitivity about 100 GeV better than CMS, while for the Gbb model the CMS H_T^{miss} analysis extends the sensitivity of the multi-b analysis by about 50 GeV.

10.2   RPV Interpretation

Throughout this thesis, only models with R-parity conservation have been considered. The analysis presented in Chap. 8 is optimized for RPC scenarios, but it maintains some sensitivity also when R-parity violating (RPV) couplings are considered. This has been studied in Ref. [10], where several ATLAS SUSY searches are reinterpreted in models with variable RPV coupling strength.

In RPC models the LSP is stable and, if it is neutral (as in the case of a $\tilde{\chi}_1^0$ LSP considered in the Gtt model), it escapes detection, giving rise to final states rich in E_T^{miss}. This is no longer the case if we allow the presence of RPV couplings: with the increase of the coupling strength, the LSP goes from being long-lived and escaping undetected, to having displaced decay vertices within the detector, to having a prompt decay.

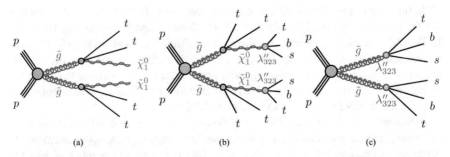

Fig. 10.3 Production and decay process for the RPV Gtt model considered. The dominant process varies with increasing λ''_{323} coupling from left to right

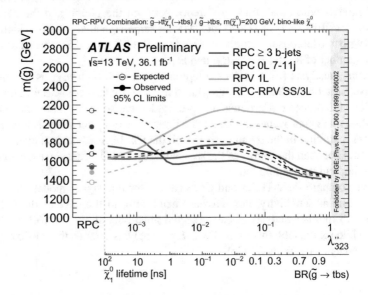

Fig. 10.4 Exclusion limits for the Gtt model as a function of λ''_{323} and $m(\tilde{g})$. Expected limits are shown with dashed lines, and observed as solid. The RPC-limit is shown on the leftmost part of the axes, while the region $\lambda''_{323} > 1.07$ is forbidden by constraints from the renormalization group equations. Figure from Ref. [10]

In particular, in the case of the Gtt model, a non-zero λ''_{323} coupling opens the decay $\tilde{\chi}^0_1 \to tbs$, and large values of λ''_{323} can also lead to the direct decay $\tilde{g} \to tbs$. The different decay modes for various values of λ''_{323} are shown in the diagrams in Fig. 10.3.

The exclusion limit from the reinterpretation of four different ATLAS searches, including the multi-b search, is shown in Fig. 10.4 in the $m(\tilde{g})$–λ''_{323} plane, assuming $m(\tilde{\chi}^0_1) = 200$ GeV and $m(\tilde{t}/\tilde{b}) = 2.4$ TeV. The multi-b search has the best expected sensitivity for a $\tilde{\chi}^0_1$ lifetime lower than 0.6 ns.

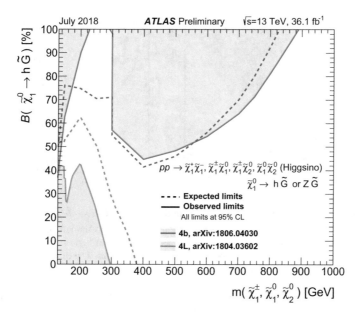

Fig. 10.5 The 95% CL exclusion limits on a general gauge mediation model from 13 TeV data. The model assumes a pure higgsino NLSP that promptly decays to either Z gravitino or Higgs gravitino. The limits are displayed as a function of the mass of the nearly mass-degenerate higgsino triplet and the branching fraction of lightest higgsino to Higgs gravitino. Figure from Ref. [2]

10.3   Higgsino Pair Production in GGM Models

The exclusion contours in the $m(\tilde{H})$–BR$(\tilde{H} \rightarrow h\tilde{G})$ plane for higgsino pair production in GGM models are shown in Figs. 10.5 and 10.6 for the ATLAS and CMS collaborations respectively.

The results included in Fig. 10.5 are the multi-b E_T^{miss}-based search described in Chap. 9, the low-mass multi-b search described in Sect. 9.10 and the four-lepton search presented in Ref. [11], that requires two pairs of same-sign opposite-flavor leptons with invariant mass compatible with a Z boson, and defines two SRs to target GMSB higgsino pair production, respectively with a 50 and 100 GeV E_T^{miss} requirement. This search is by construction more sensitive to models with a high BR$(\tilde{H} \rightarrow Z\tilde{G})$.

The CMS Collaboration has a more extensive program covering this signal model for what concerns the analysis of the 2015–2016 dataset. Different final states are used to be sensitive to the possible combinations of branching ratios, providing a good sensitivity throughout the full plane. Figure 10.6a shows the combined exclusion contour: the expected sensitivity is up to about 580 GeV in higgsino mass in the case where BR$(\tilde{H} \rightarrow Z\tilde{G}) = 100\%$ and 800 GeV for BR$(\tilde{H} \rightarrow h\tilde{G}) = 100\%$.

Figure 10.6b shows the observed exclusion limit for each individual analysis. We can see that, as it is the case for ATLAS, the 4b search [13] provides the best sensitivity

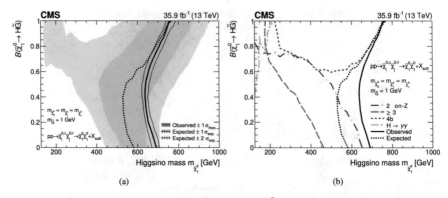

(a)   (b)

Fig. 10.6 Exclusion contours at the 95% CL in the $m(\tilde{\chi}_1^0)$–$BR(\tilde{H} \to h\tilde{G})$ plane for the model of $\tilde{\chi}_1^0 \tilde{\chi}_1^0$ production. **a** Combined exclusion contours. The area to the left of or below the solid (dashed) black curve represents the observed (expected) exclusion region. The green and yellow bands indicate the $\pm 1\sigma$ and 2σ uncertainties in the expected limit. The thin black lines show the effect of the theoretical uncertainties ($\pm 1\sigma_{theory}$) on the signal cross-section. **b** Observed contours for each individual CMS analysis compared with the combination. For the 4b contour, the region above is excluded, while for all others, the region to the left is excluded. The 4b search drives the exclusion at large values of $BR(\tilde{H} \to h\tilde{G})$ while the on-Z dilepton and multilepton searches are competing at lower values of $BR(\tilde{H} \to h\tilde{G})$. Figures from Ref. [12]

for models with high $BR(\tilde{H} \to h\tilde{G})$, except for low $m(\tilde{H})$, where it is instead the two-photon search [14] that provides the most stringent limit. This is because the 4b analysis in CMS selects events that fire one of multiple triggers, all of which require some E_T^{miss}. Instead the ATLAS analysis includes also the dedicated search optimized for low-mass signal. The signals with intermediate and high $BR(\tilde{H} \to Z\tilde{G})$ are probed with a search with two leptons consistent with the Z-boson peak [15]. ATLAS at the moment does not have a two-lepton search that provides a good sensitivity to this model, and therefore it has overall less coverage in the $m(\tilde{H})$–$BR(\tilde{H} \to h\tilde{G})$ plane. Despite this, the comparison of the CMS 4b search with the multi-b search from ATLAS described in Chap. 9 shows a better sensitivity of the ATLAS search, especially at high higgsino mass, where the very low-background SRs of the ATLAS analysis provide better sensitivity.

10.4   ATLAS Mass Reach

Figure 10.7 shows the mass reach of the ATLAS SUSY searches on some representative signal models, as of July 2018.

Signal models are divided by production mode and/or model assumptions into:

Inclusive searches   Searches that look for direct production of squarks (except 3rd generation squarks) and gluinos, including the gluino multi-b analysis, are grouped into this category. For the multi-b analysis, only the Gtt result is reported.

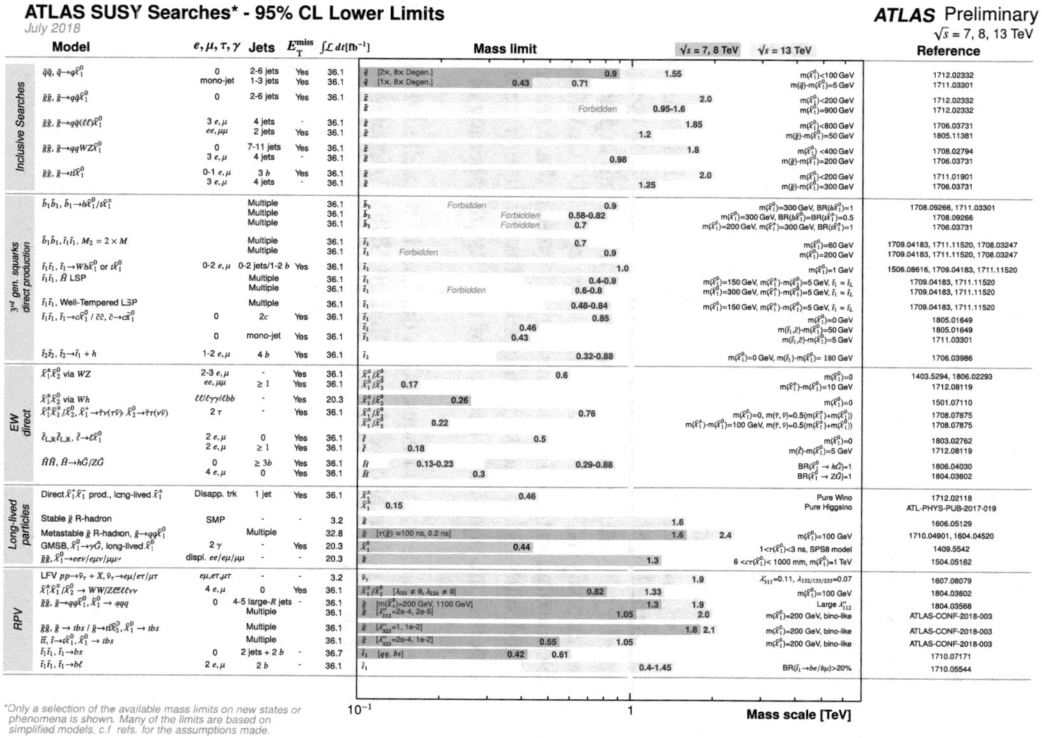

ATLAS SUSY Searches* · 95% CL Lower Limits
July 2018

ATLAS Preliminary
$\sqrt{s} = 7, 8, 13$ TeV

Fig. 10.7 Mass reach of the ATLAS searches for Supersymmetry. A representative selection of the available search results is shown. Results are quoted for the nominal cross-section in both: a region of near-maximal mass reach and a demonstrative alternative scenario, in order to display the range in model space of search sensitivity. Some limits depend on additional assumptions on the mass of the intermediate states, as described in the references provided in the plot. Figure from Ref. [2]

The assumptions of this model lead to a spectacular final state, rich in jets and b-jets, that allows us to effectively separate signal from background, leading to the gluino mass limit of 2 TeV for low neutralino mass (note that the multi-b limit reported in Fig. 10.7 is with the 2015–2016 data only). This should not be translated into the universal statement that gluinos with mass of 2 TeV are excluded: as it has already been noticed in Chap. 8, a higher neutralino mass weakens substantially the observed limit. From Fig. 10.7 we can further learn that the assumptions on the gluino decay mode can alter the limits as well, and in fact all the limits are weaker than the ones we obtain for decays to third generation particles.

3rd generation squark direct production   In natural SUSY models third generation squarks are expected to be relatively light, and this motivates the extensive ATLAS program to search for stop and sbottom production. Also in this case, Fig. 10.7 shows how the limits set by the different searches depend a lot on the decay mode assumed; for example, in the case of $m(\tilde{t}_1)$, the observed limits range from 430 GeV in the case of a decay to the charm quark to up to 1 TeV if the decay is instead to a top quark.

EW direct   The direct production of charginos, neutralinos and sleptons has the lowest limits due to the lower electroweak production cross-section compared to processes mediated by the strong force. The strongest mass limit is placed by the multi-b electroweak analysis, that excludes Higgisnos up to 880 GeV. The wider exclusion range compared to the other electroweak analyses is due to the peculiar features of the final state and to the large higgsino cross-section (four different production modes that are equivalent because of the mass degeneracy of $\tilde{\chi}_1^0$, $\tilde{\chi}_2^0$, $\tilde{\chi}_1^\pm$).

Long-lived particles   The analyses discussed in this thesis, as well as the models mentioned in the three points above, assume that all SUSY particles originate a chain of prompt decays to SM particles and the LSP. Including in the signal models a non-zero lifetime for SUSY particles leads to very different signatures that can be exploited to search for BSM physics. In this case the range of exclusion depends on the lifetime itself, and in favorable cases it can lead to stronger bounds than the ones obtained for prompt decays; this is the case for example for gluinos decaying to light quarks, that are excluded up to 2.4 TeV for a gluino lifetime of 0.2 ns.

RPV   The reason why a large number of the ATLAS SUSY searches require the presence of E_T^{miss} is that in R-parity conserving models the LSP is long lived and therefore, if it has zero charge, it escapes detection. In RPV models this is no longer the case, and the final states have less E_T^{miss} (originating only from neutrinos form SM decays) and in general more visible decay products. The limits placed in the case of RPV models are similar or stronger than those placed for corresponding RPC scenarios, as it is the case e.g. for the $\tilde{t}_1$, excluded between 400 GeV and 1.45 TeV if the $B - L$-conserving decay $\tilde{t}_1 \rightarrow bl$ is allowed. As already discussed in Sect. 10.2, the strength of the RPV couplings can change noticeably the sensitivity of the analyses.

References

1. ATLAS Collaboration (2017) Search for supersymmetry in final states with two same-sign or three leptons and jets using 36 fb^{-1} of $\sqrt{s} = 13$ TeV pp collision data with the ATLAS detector. JHEP 09:084. arXiv:1706.03731 [hep-ex]
2. ATLAS Collaboration (2018) https://twiki.cern.ch/twiki/bin/view/AtlasPublic/ SupersymmetryPublicResults. Accessed 2018-07-25
3. ATLAS Collaboration (2017) Search for new phenomena with large jet multiplicities and missing transverse momentum using large-radius jets and flavour-tagging at ATLAS in 13 TeV pp collisions. JHEP 12:034. arXiv:1708.02794 [hep-ex]
4. CMS Collaboration (2017) Search for supersymmetry in multijet events with missing transverse momentum in proton-proton collisions at 13 TeV. Phys Rev D 96:032003. arXiv:1704.07781 [hep-ex]
5. CMS Collaboration (2018) Search for supersymmetry in proton-proton collisions at 13 TeV using identified top quarks. Phys Rev D 97:012007. arXiv:1710.11188 [hep-ex]
6. CMS Collaboration (2017) Search for physics beyond the standard model in events with two leptons of same sign, missing transverse momentum, and jets in proton-proton collisions at $\sqrt{s} = 13$ TeV. Eur Phys J C 77:578. arXiv:1704.07323 [hep-ex]
7. CMS Collaboration (2017) Search for new phenomena with the M_{T2} variable in the all-hadronic final state produced in proton-proton collisions at $\sqrt{s} = 13$ TeV. Eur Phys J C 77:710. arXiv:1705.04650 [hep-ex]
8. CMS Collaboration (2018) Search for natural and split supersymmetry in proton-proton collisions at $\sqrt{s} = 13$ TeV in final states with jets and missing transverse momentum. JHEP 05:025. arXiv:1802.02110 [hep-ex]
9. CMS Collaboration (2018) https://twiki.cern.ch/twiki/bin/view/CMSPublic/PhysicsResults SUS. Accessed 2018-08-01
10. ATLAS Collaboration (2018) Reinterpretation of searches for supersymmetry in models with variable R-parity-violating coupling strength and long-lived R-hadrons. ATLAS-CONF-2018-003. https://cds.cern.ch/record/2308391
11. ATLAS Collaboration. Search for supersymmetry in events with four or more leptons in $\sqrt{s} = 13$ TeV pp collisions with ATLAS. arXiv:1804.03602 [hep-ex]
12. CMS Collaboration (2018) Combined search for electroweak production of charginos and neutralinos in proton-proton collisions at $\sqrt{s} = 13$ TeV. JHEP 03:160. arXiv:1801.03957 [hep-ex]
13. CMS Collaboration (2018) Search for Higgsino pair production in pp collisions at $\sqrt{s} = 13$ TeV in final states with large missing transverse momentum and two Higg bosons decaying via $H \rightarrow b\bar{b}$. Phys Rev D 97:032007. arXiv:1709.04896 [hep-ex]
14. CMS Collaboration (2018) Search for supersymmetry with Higgs boson to diphoton decays using the razor variables at $\sqrt{s} = 13$ TeV. Phys Lett B 779:166. arXiv:1709.00384 [hep-ex]
15. CMS Collaboration (2018) Search for new phenomena in final states with two opposite-charge, same-flavor leptons, jets, and missing transverse momentum in pp collisions at $\sqrt{s} = 13$ TeV. JHEP 03:076. arXiv:1709.08908 [hep-ex]

Chapter 11
Conclusion

This dissertation presented two searches targeting SUSY signals leading to final states with high b-jet multiplicity, using the data collected by the ATLAS experiment at the LHC between 2015 and 2016, at a center-of-mass energy $\sqrt{s} = 13$ TeV. This dataset corresponds to an integrated luminosity of 36.1 fb^{-1}.

The first analysis searches for gluino pair production where each gluino decays through a stop or a sbottom to respectively four top or bottom quarks and the LSP, leading to a final state with four b-jets and missing transverse momentum ($E_\mathrm{T}^\mathrm{miss}$). This analysis employs two different strategies: cut-and-count, with several non-orthogonal SRs optimized to maximize the discovery significance for selected benchmark models, and multi-bin, with orthogonal SRs that are statistically combined in the maximum-likelihood fit to maximize the exclusion power of the analysis. In all the SRs semi-leptonic $t\bar{t}$ +jets constitutes the dominant background. No significant excess is found in any of the analysis regions. The largest deviation between expected and observed number of events is in SR-0L-HH, one of the multi-bin SRs, and it has a significance of approximately 2.3 standard deviations. Exclusion limits in the m($\tilde{g}$)-m($\tilde{\chi}_1^0$) are set for the two simplified models assuming 100% BR for the gluino into $t\bar{t}\tilde{\chi}_1^0$ and $b\bar{b}\tilde{\chi}_1^0$, denoted as "Gtt" and "Gbb" respectively. In the case of the Gtt model, the expected and observed limit at 95% CL for massless neutralino are 2.14 and 1.97 TeV respectively; the main reason of the difference is the excess in SR-0L-HH. Also for the Gbb model the observed limit is slightly weaker than the expected one: while the expected limit for massless neutralino is 2.01 TeV, the observed is 1.92 TeV. The results of this analysis are reinterpreted also allowing a variable BR of the gluino into tt$\tilde{\chi}_1^0$, bb$\tilde{\chi}_1^0$ and tb$\tilde{\chi}_1^\pm$. These limits, as well as the other limits discussed in this thesis, are obtained with the CL$_s$ prescription. These results are published in Ref. [1].

© Springer Nature Switzerland AG 2020
C. Rizzi, *Searches for Supersymmetric Particles in Final States*
with Multiple Top and Bottom Quarks with the Atlas Detector, Springer Theses,
https://doi.org/10.1007/978-3-030-52877-5_11

The mild excess in SR-0L-HH has been verified also with the 2017 dataset, leaving the definition of the SRs unchanged; together with the 2015 and 2016 data-taking periods, this leads to a total integrated luminosity of 79.8 fb^{-1}. The update of the analysis did not find an excess in SR-0L-HH, and the increase in luminosity allowed to set more stringent limits on gluino pair production. In the case of the Gtt model, for neutralino masses below 800 GeV we excluded gluino masses up to 2.25 TeV, while for the Gbb model the limit is at 2.17 TeV. The results of the analysis of the 79.8 fb^{-1} are reinterpreted also for signals with on-shell stops. In this case it is found that, while for most m($\tilde{t}$) the sensitivity is close to the one obtained in the off-shell case, it becomes weaker when m($\tilde{t}$) is close to the kinematic boundary of m($\tilde{\chi}_1^0$)+m(t) and m($\tilde{g}$)-m(t). These results have been released in Ref. [2].

The second search presented in this thesis targets a GGM model of higgsino pair production, where each higgsino then decays to a Higgs boson and a gravitino, which in this case is the LSP. The search is performed in the channel with four b-jets, originating from the decay of the two Higgs bosons, and E_T^{miss}. This was the first ATLAS analysis targeting this signature, that had been previously considered only in searches performed by the CMS collaboration. This analysis relies on the identification of two Higgs boson candidates in events with at least three or at least four b-tagged jets. Several orthogonal SRs are optimized and statistically combined in the fit. Two discovery SRs are also defined to provide stronger model-independent limits. If we assume that the higgsino decays to Higgs boson and gravitino with 100% BR, this analysis excludes $m(\tilde{H})$ in the range 240–880 GeV at 95% CL. This analysis is complemented by a second analysis targeting signals with low $m(\tilde{H})$, where the invisible momentum in the final state is not enough to fire the E_T^{miss} trigger. Because of a mild excess in this latter analysis, the excluded range of higgsino masses is between 130 and 230 GeV and between 290 and 880 GeV. The results are also interpreted in models with a variable BR of the higgsino into Higgs or Z boson. For $m(\tilde{H}) = 400$ GeV, signal models with BR to Higgs boson higher than 45% are excluded at 95% CL. These results are published in Ref. [3].

The results presented in this thesis are an important element in the wide ATLAS program for SUSY searches: they provide some of the most restrictive bounds on Natural SUSY scenarios, and the experience gained in developing them represents a stepping stone to more sensitive searches with the data to be collected in the coming years. So far no significant deviation from the SM predictions has been found, but the LHC is still taking data, and the increase in luminosity as well the effort in constantly improving the analysis techniques will allow to improve the sensitivity and probe also models that could have escaped detection in previous analyses. It should also always be kept in mind that all the model-dependent limits we present are based on simplified models with strong assumptions in terms of reachable particles and possible decay chains, and they typically become much weaker once we recast the existing analysis to more realistic models. The LHC Run 2 will continue its pp-collision program until autumn 2018, providing an expected integrated luminosity of 140 fb^{-1}; after this it will undergo a two-year-long shutdown, to resume operations in 2021 for three years of data taking at $\sqrt{s} = 14$ TeV, with 2.5 times the nominal luminosity. During this period the size of the dataset collected by ATLAS is expected to reach a total of approximately 300 fb^{-1}, providing an unprecedented opportunity to explore the energy frontier and possibly leading to fascinating breakthroughs in our understanding of physics beyond the SM.

References

1. ATLAS Collaboration (2018) Search for supersymmetry in final states with missing transverse momentum and multiple b-jets in proton-proton collisions at $\sqrt{s} = 13$ TeV with the ATLAS detector. JHEP 06:107. arXiv:1711.01901 [hep-ex]
2. ATLAS Collaboration (2018) Search for supersymmetry in final states with missing transverse momentum and multiple b-jets in proton-proton collisions at $\sqrt{s} = 13$ TeV with the ATLAS detector. ATLAS-CONF-2018-041. http://cds.cern.ch/record/2632347
3. ATLAS Collaboration (2018) Search for pair production of higgsinos in final states with at least three b-tagged jets in $\sqrt{s} = 13$ TeV pp collisions using the ATLAS detector, Submitted to: Phys. Rev. arXiv:1806.04030 [hep-ex]

Appendix A
Kinematic Reweighting in the Gluino Search

In this appendix we discuss in more details the kinematic reweighting mentioned in Sect. 8.5, designed to mitigate the effect of the mismodeling that affects all the energy-related variables in the 1-lepton channel, where we observe a downward trend in the data/MC ratio, clearly visible e.g. in the bottom panel of Fig. 8.29a; this does not happen instead in the 0-lepton channel, as it can be observed in Fig. 8.29b. This difference in trends is problematic for the analysis, since all of the VRs require at least one signal lepton, including the ones used to derive the prediction for 0-lepton SRs.

To have a better estimate of the background in the high-m_{eff} regime, a reweighting has been derived to bring the MC prediction closer to the observed data. This reweighting is computed in a region that requires:

- at least 4 jets,
- exactly 2 b-jets,
- $E_{\text{T}}^{\text{miss}} > 200$ GeV,
- at least one lepton,
- $m_{\text{T,min}}^{b\text{-jets}} < 140$ GeV.

This selection is orthogonal to SRs and CRs but has a similar background composition. The $m_{\text{T,min}}^{b\text{-jets}}$ selection reduces further the signal contamination and allows for a validation region with exactly 2 b-jets and high $m_{\text{T,min}}^{b\text{-jets}}$. The same trend in the data/MC ratio in the m_{eff} distribution has been observed also in other regions non dominated by the $t\bar{t}$ background, therefore the reweighting is computed taking into account all backgrounds and applied to all backgrounds. The binning of the reweighting is chosen to be 50 GeV for most of the m_{eff} spectrum, while at high m_{eff} the width of the bins is larger to allow ≈ 100 data events in each bin. Before deriving the reweighting, the sum of the MC predictions for the different backgrounds is scaled to the same yield as in data, so that the reweighting. The final reweighting for the 1-lepton channel is shown in the bottom panel of Fig. A.1.

© Springer Nature Switzerland AG 2020
C. Rizzi, *Searches for Supersymmetric Particles in Final States*
with Multiple Top and Bottom Quarks with the Atlas Detector, Springer Theses,
https://doi.org/10.1007/978-3-030-52877-5

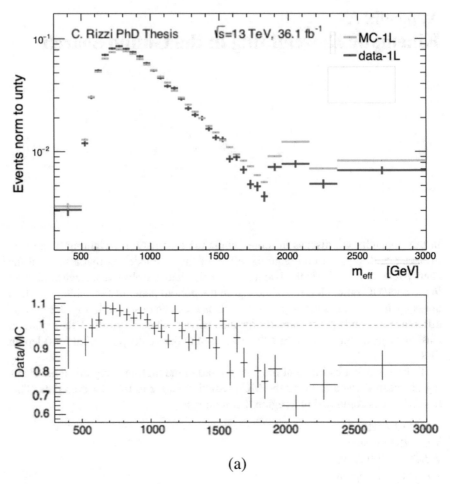

Fig. A.1 m_{eff} distribution in data and simulation in a selection that requires at least 4 jets, exactly 2 b-jets, $E_{\text{T}}^{\text{miss}} > 200$ GeV, at least one lepton, and $m_{\text{T,min}}^{b\text{-jets}} < 140$ GeV

A.1 Effect on other variables

Figures A.2, A.3 and A.4 show the effect of the reweighting on the main kinematic variables. We can see that the reweighting, derived using only the m_{eff} distribution, improves the comparison between data and the simulation in most of the energy-related variables.

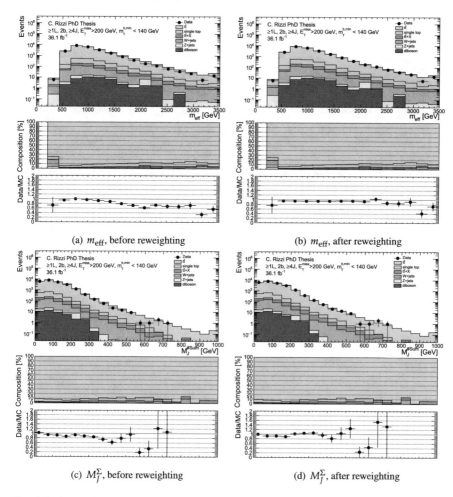

(a) m_{eff}, before reweighting

(b) m_{eff}, after reweighting

(c) M_J^Σ, before reweighting

(d) M_J^Σ, after reweighting

Fig. A.2 Comparison between data and simulation in the reweighting region, before and after applying the kinematic reweighting

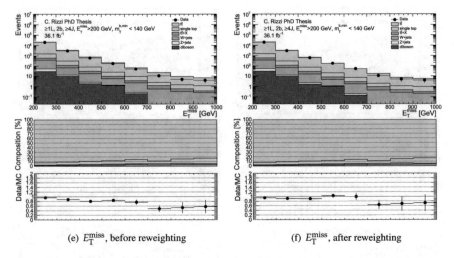

(e) E_T^{miss}, before reweighting   (f) E_T^{miss}, after reweighting

Fig. A.2 (continued)

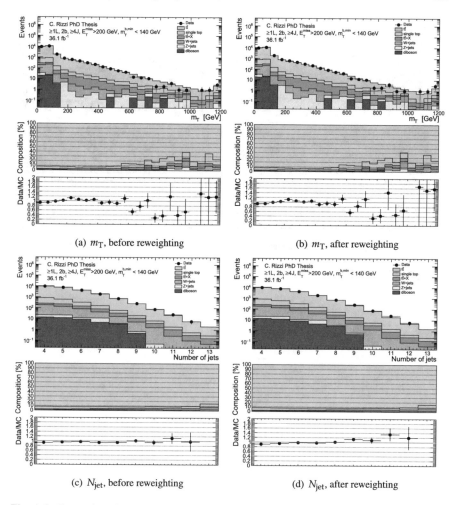

(a) m_T, before reweighting

(b) m_T, after reweighting

(c) N_{jet}, before reweighting

(d) N_{jet}, after reweighting

Fig. A.3 Comparison between data and simulation in the reweighting region, before and after applying the kinematic reweighting

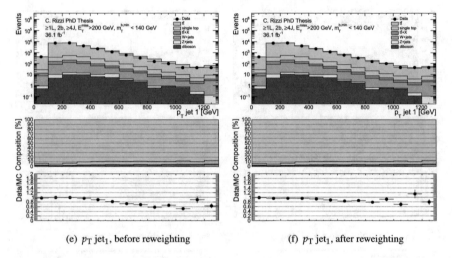

(e) p_T jet$_1$, before reweighting   (f) p_T jet$_1$, after reweighting

Fig. A.3   (continued)

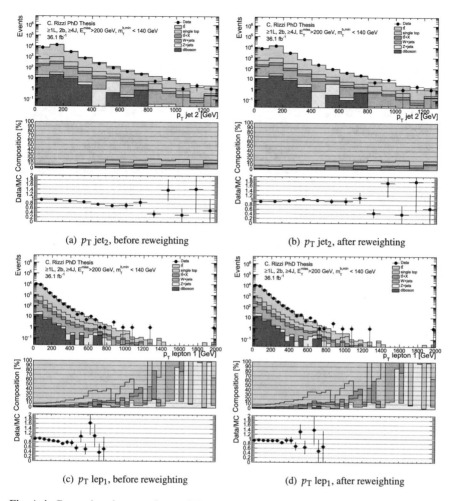

(a) p_T jet$_2$, before reweighting

(b) p_T jet$_2$, after reweighting

(c) p_T lep$_1$, before reweighting

(d) p_T lep$_1$, after reweighting

Fig. A.4 Comparison between data and simulation in the reweighting region, after applying the kinematic reweighting

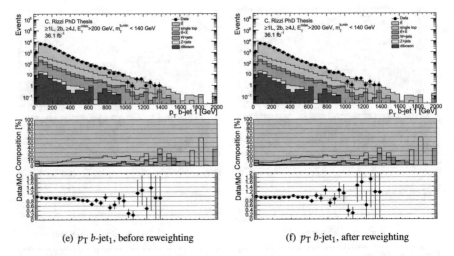

(e) p_T b-jet$_1$, before reweighting   (f) p_T b-jet$_1$, after reweighting

Fig. A.4 (continued)

Appendix B
Higgs Boson Reconstruction in the Higgsino Search

This appendix discusses the approaches that are tested for the reconstruction of the candidate Higgs bosons for the higgsino search described in Chap. 9. The signal events considered in this appendix are those where both Higgs bosons decay to a $b\bar{b}$ pair. As already discussed in Sect. 9.2, the four jets selected to reconstruct the two Higgs bosons are selected with the following criteria:

- If there are exactly four b-tagged jets in the event, those are used.
- If there are more than four b-tagged jets, the selected ones are the four b-tagged jets with highest p_T.
- If there are less than four b-tagged jets, the selected ones are the b-tagged jets and the non-tagged jets with highest p_T.

Figure B.1 shows the fraction of signal events that have four reconstructed jets, the fraction of signal events where it is possible to select the four correct jets originating from the decay of the Higgs bosons (which corresponds to requiring four jets with $p_T > 25$ GeV matched in dR < 0.3 with the 4 b-quarks originating from the two Higgs bosons), and the fraction of signal events where the choice of jets described above selects the correct jets. The four jets selected are the correct set of jets about 70–80% of the times that the correct match is possible.

Once the four jets are selected, different algorithms to group them into pairs (each one corresponding to one of the two Higgs boson candidates) are compared:

min-diff Minimize the difference between m(h_1) and m(h_2), where m(h_1) and m(h_2) are the masses of the two boson candidates and m(h_1)>m(h_2).
min-dR Minimize ΔR^{bb}_{max} (defined in Sect. 9.2).
max-p_T Maximize min(p_T (h_1), p_T (h_2)).

Figure B.2 shows the fraction of times that each of the algorithms described above leads to the same pairs as the true matching, with respect to the number of events in which the true matching is possible. For signals with low higgsino mass the algorithm that minimize the mass difference performs better, while at high signal

© Springer Nature Switzerland AG 2020
C. Rizzi, *Searches for Supersymmetric Particles in Final States
with Multiple Top and Bottom Quarks with the Atlas Detector*, Springer Theses,
https://doi.org/10.1007/978-3-030-52877-5

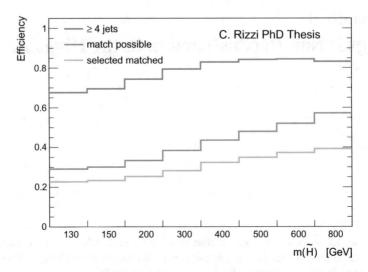

Fig. B.1 Fraction of events where the algorithm described in the text selects the correct jets

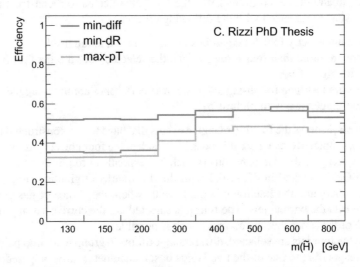

Fig. B.2 Fraction of hh→4b events where the reconstruction method indicated in the legend leads to the correct match. The fraction shown is with respect to the events where the correct match is possible

masses, where the two Higgs bosons (and their decay products) are more boosted, the min-dR algorithm reproduces the true matching a higher fraction of times. The max-p_T algorithm instead underperforms for all signal masses compared to the other two.

Figures B.3 and B.4 show the distribution of $m(h_1)$ and $m(h_2)$ respectively, for signal and background events normalized to unit area, with Higgs boson candidates reconstructed with the min-dR and min-diff algorithms. We can notice the different

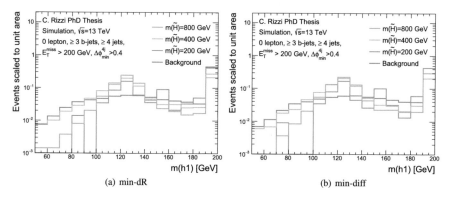

Fig. B.3 Distribution of m(h_1) in signal and background. The Higgs candidates are reconstructed with **a** the min-dR algorithm and **b** the min-diff algorithm. All distributions are normalized to unit area

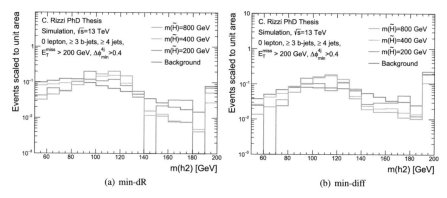

Fig. B.4 Distribution of m(h_2) in signal and background. The Higgs candidates are reconstructed with **a** the min-dR algorithm and **b** the min-diff algorithm. All distributions are normalized to unit area

shape of the distributions with the two algorithms, especially in the case of m(h_2). Considering that the high-mass analysis focuses on signals with intermediate and high higgsino mass, the min-dR algorithm is chosen as baseline algorithm for the reconstruction of the candidate Higgs bosons. This choice is confirmed by the optimization procedure described in Sect. 9.3.1: if the values of the candidate Higgs bosons reconstructed with both algorithms have are as input, the min-dR algorithm gives consistently better expected significance.

Appendix C
Comparison Between Data and Simulation for the Higgsino Search

Figures C.1 and C.2 contain the data-MC comparison for the analysis variables not shown in Chap. 9.

© Springer Nature Switzerland AG 2020
C. Rizzi, *Searches for Supersymmetric Particles in Final States with Multiple Top and Bottom Quarks with the Atlas Detector*, Springer Theses,
https://doi.org/10.1007/978-3-030-52877-5

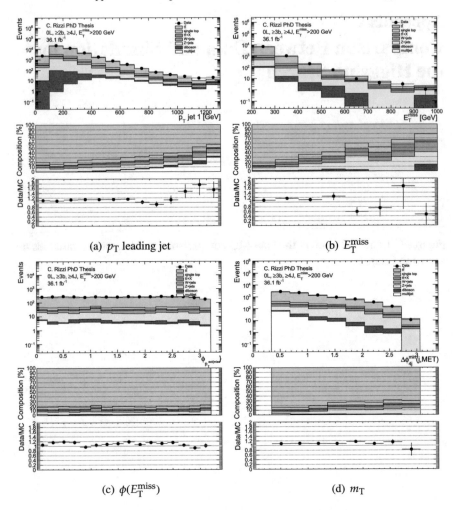

(a) p_T leading jet

(b) E_T^{miss}

(c) $\phi(E_T^{miss})$

(d) m_T

Fig. C.1 Comparison between data and simulation in the 0-lepton channel

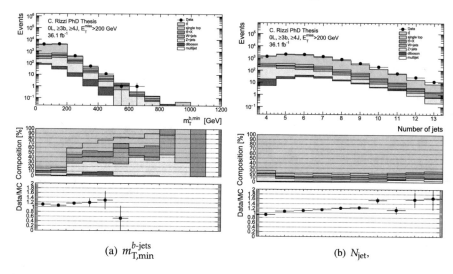

(a) $m_{\mathrm{T,min}}^{b\text{-jets}}$

(b) N_{jet},

Fig. C.2 Comparison between data and simulation in the 0-lepton channel

Appendix D
TileCal PMT Response in Calibration Transfer Analysis

This appendix describes the impact of a non linearity in the response of the TileCal PMTs on the luminosity calibration transfer analysis; this is used to estimate the uncertainty in the luminosity due to the extrapolation between low-μ bunches, where LUCID is calibrated, and the high-μ bunches of physics runs. Correcting for this non-linearity in the PMTs allows to considerably reduce the luminosity systematic uncertainty.

D.1 Luminosity Calibration Transfer

The general strategy to measure the luminosity in ATLAS is outlined in Sect. 3.3.6. LUCID is the ATLAS detector that, during vdM runs, measures the absolute luminosity. LUCID algorithms are non-linear with μ, and this non-linearity is corrected with the calibration transfer, which allows to extrapolate the absolute LUCID calibration from conditions corresponding to few low-μ, isolated bunches to the conditions of the physics runs, where there are many high-μ bunches in trains.

The default system used to provide the calibration transfer is the tracking system. As already discussed in Sect. 3.3.6, the number of reconstructed tracks in the ID is proportional to μ. The track selection that is used for the calibration transfer corresponds to the TightPrimary quality criteria but dropping the requirement on the Pixel holes (that have to be ≤ 1 for the standard TightPrimary selection); furthermore there is a requirement on the impact parameter significance, $|d_0|/\sigma(d_0) < 7$, and on the pseudorapidity of the track, $|\eta| < 1$. These criteria have been optimized to reduce the dependence on the Pixel conditions.

The procedure to determine the calibration transfer assumes that the tracking luminosity measurement (in the following referred to as "Tracking") does not have any dependence on μ or on the position in the train, and it consists of three steps:

© Springer Nature Switzerland AG 2020
C. Rizzi, *Searches for Supersymmetric Particles in Final States with Multiple Top and Bottom Quarks with the Atlas Detector*, Springer Theses,
https://doi.org/10.1007/978-3-030-52877-5

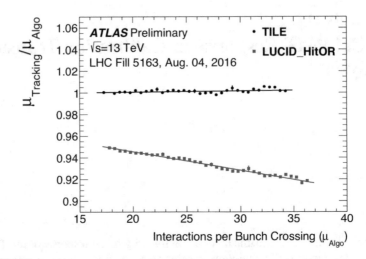

Fig. D.1 Ratio of the Tracking luminosity to the luminosity measured by LUCID (red) and TileCal (black) as a function of μ, in LHC physics fill 5163 with 2064 colliding bunches in 2016. Figure from Ref. [1]

1. The bunch-averaged track luminosity during the vdM run is normalized to the LUCID luminosity (which in these runs undergo the absolute calibration). The vdM runs have low-μ isolated bunches.
2. The ratio of the luminosity obtained with LUCID with Tracking is measured as a function of μ (as measured by LUCID) in a single run with high-μ trains. The distribution of the values of this ratio is fitted with a straight line.
3. The result of the fit is used to apply a correction to the luminosity measured by LUCID in physics runs.

As it can be observed in Fig. D.1, at high-μ LUCID overestimates the luminosity and the correction factor is as big as 11% for $\mu \approx 50$.

To assign an uncertainty to the calibration transfer, a procedure similar to that used for the Tracking luminosity is repeated using the integrator system of the TileCal detector, and the relative difference between the luminosity measured by the tracking system and by TileCal is used as uncertainty. Just like in the case of the luminosity measurement from the tracking system, the TileCal luminosity measurement relies on some assumptions, in particular the perfect linearity of the PMT response over a wide range of instantaneous luminosity, which spans over three orders of magnitude.

D.2　TileCal Laser System

The TileCal laser system [2] is designed with the main purpose of calibrating the PMTs and readout chain: in absence of collisions, a controlled amount of light is sent to each PMT's photocatode, and the response is used to derive the laser calibration constant. The laser system has been renewed for Run 2 and the new version (LaserII) has been installed in October 2014 [3]. During the calibration, which is performed every two or three days in the pauses between the LHC collisions, laser pulses with a wavelength of 532 nm are sent to all PMT cathodes through 400 100-m long fibers. The laser light is sent also to monitor photodiodes, to remove the dependence on the laser stability.

Laser pulses are also sent in the abort gaps during physics runs, with a frequency of 3 Hz. This procedure is used to detect "time jumps", changes in the time settings of groups of channels that in Run 1 were particularly frequent after a power restart of the low-voltage power supply. The response of the PMTs to the laser pulses sent in empty bunches in physics runs is also used to perform the analysis described in this appendix.

D.3 PMT Response to Laser Pulses in Empty Bunches

One of the possible techniques to estimate the difference between the Tracking and the TileCal measurements of the calibration transfer is the following:

1. The luminosity of all the TileCal cells used is individually calibrated to match the Tracking luminosity at a specific high-μ run (anchor run).
2. Interpolating this value with the origin in the current–luminosity plane allows to derive an estimate for the TileCal luminosity during the vdM run.
3. The comparison of this luminosity with the luminosity measured by the tracking system provides the systematic uncertainty in the calibration transfer.

The measurement of the calibration transfer with TileCal needs to take into account two effects that complicate the measurement.

- Activation decays after high-μ runs bias the TileCal response in low-μ runs like the vdM run.
- It has been shown that the PMT response is not perfectly linear with the luminosity.

This sections focus on the analysis of this second point and on the corrections derived to minimize its effect on the TileCal calibration transfer analysis; this allows to reduce the calibration transfer uncertainty in the 2017 luminosity, which is the dominant luminosity uncertainty for the 2016 dataset. The cells we consider for this study are all the E-type cells and A13 (see Fig. 3.12).

It has previously been noticed (see e.g. Ref. [4]) that the TileCal PMTs show a non-linearity in the response with the increase in luminosity. In this section we study this effect for the cells and run numbers that are of interest for the calibration transfer analysis. The anchor run used in the calibration transfer analysis is 331085, while the run number of the vdM scan is 330875.

The TileCal laser system is primarily used to calibrate the TileCal readout. The laser system can also be fired in the abort gaps during standard physics runs; in this case one laser pulse is sent three times per second. The PMT response is analyzed in the following steps:

1. The response of the PMTs is normalized to the response of the monitor diode D0 to remove fluctuations due to laser instabilities.
2. The distribution of the response for each individual PMT in groups of 25 LB is considered. Grouping together several LBs is necessary to accumulate enough data points: the frequency of laser pulses is 3 Hz, which gives only about 180 data points per minute. The value of 25 has been chosen for consistency with previous studies, after checking that the size of the group of LBs does not affect the results as long as it is large enough to provide a sufficiently large number of events.
3. The response for each cell family is computed by averaging over all the PMTs belonging to that family. We keep separate the left and right PMTs and the A and C side of TileCal.
4. The distribution of the response is normalized to the last group of LBs that does not contain the LB where "stable beams" is declared, which is used as reference.

Figure D.2 shows the PMT response for the cell family A13 during the anchor run, while Fig. D.3 shows the response for the cells of the families E1 and E2 and Fig. D.4 for the families E3 and E4. While the origin of this discontinuity in the response is still under investigation, it is clear that the drastic change in response happens in correspondence of the declaration of stable beam, when the luminosity increases, and is therefore referred to as non-linearity of the PMTs.

The PMT non-linearity is quantified in three different ways, which in the Figures are labeled as:

Jump   Relative difference between the first group of LBs after "stable beams" declaration that does not contain the LB where "stable beams" is declared and the group of LBs used as reference.

Average   Average of the response after "stable beams" is declared until the end of the run, relative to the reference.

Lumi-average   As above, but the average is weighted by the amount of luminosity collected in each group of LBs.

Equivalent studies on run 330875 show that for the low luminosity of the vdM run there is no discontinuity in the distribution of the PMT response, and therefore no laser correction is needed for the vdM run.

D.4 Impact on Calibration Transfer Uncertainty

The change in the PMT response with the increase in luminosity has a direct implication in the TileCal luminosity measurement. In particular, it means that the increase in measured current with the increase in luminosity comes from two distinct factors:

- The actual increase in luminosity, i.e. having more particles traversing the detector.

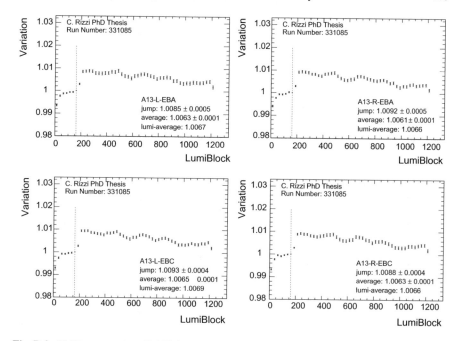

Fig. D.2 PMT response in cell A13 for the anchor run (run number 331085). The dashed red line indicates the last group of LBs that does not contain the LB where "stable beams" is declared, which is used as reference

- The increase in PMT response.

While the first bullet is the effect that we want to measure to provide a luminosity calibration, the second bullet has the effect of artificially increasing the TileCal luminosity measurement. The value of the PMT non-linearity measured in Sect. D.3 is used to correct for this undesired effect, with the net result of TileCal providing a higher value for the luminosity measurement for the vdM run, as schematically illustrated in Fig. D.5.

Figures D.6 and D.7 show the effect of the correction derived for the cell families used in the computation of the calibration transfer uncertainty, for the A side and C side respectively.

To quantify the improvement provided by the laser corrections, we can compute the integrated luminosity over the whole vdM run for the TileCal cell families and compare it to the integrated luminosity from the tracking system; this is shown in Fig. D.8. The average relative difference between TileCal and Tracking luminosity, averaged over all the considered cell families, results to be:

- no PMT correction: 2.19%,
- jump: 1.19%,
- average: 1.32%,
- Lumi-average: 1.25%.

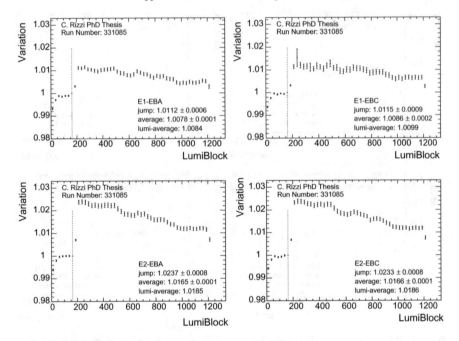

Fig. D.3 PMT response in cell E1 and E2 for the anchor run (run number 331085). The dashed red line indicates the last group of LBs that does not contain the LB where "stable beams" is declared, which is used as reference

We can see that the inclusion of the laser corrections reduces the uncertainty by about 1% absolute.

The calibration transfer uncertainty for the 2017 data-taking period used for the first public results with the 2017 dataset is computed with a similar procedure but with some small differences. In particular:

- The luminosity is re-anchored at the vdM run, and the difference is evaluated at run 331085.
- Only E-type cell families are considered.
- The anchoring is done considering only the first half of the vdM run, where the pedestal subtraction is more reliable.

The numerical value for the laser correction used is labeled as "average" in Figs. D.2, D.3 and D.4. Despite these differences, the uncertainty resulting from the average of the relative difference between TileCal and Tracking luminosities for the different cell families is 1.3%. Also in this case, ignoring the laser correction derived in Sect. D.3 would lead to an average of the differences of about 1% higher.

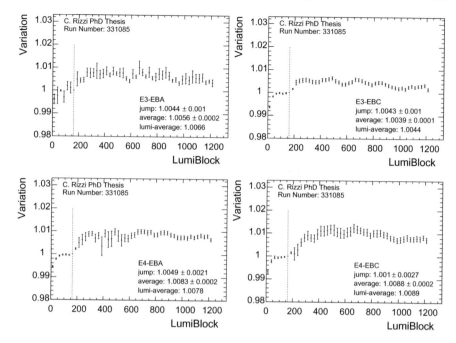

Fig. D.4 PMT response in cell E3 and E4 for the anchor run (run number 331085). The dashed red line indicates the last group of LBs that does not contain the LB where "stable beams" is declared, which is used as reference

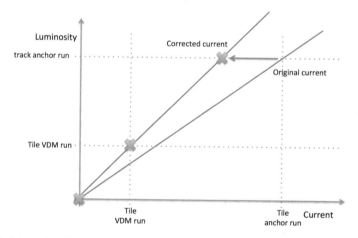

Fig. D.5 Schematic effect of the impact of the correction of the PMT response on the TileCal luminosity measurement

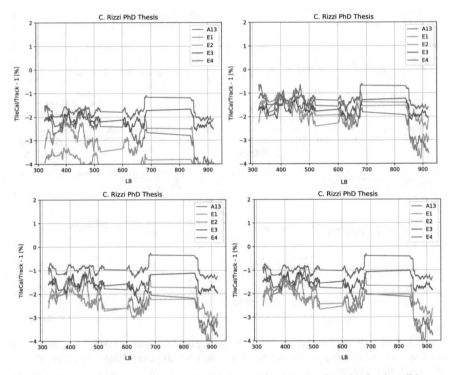

Fig. D.6 Fractional difference between the TileCal and Tracking luminosities for the vdM run **a** without any laser correction, **b** with the "jump" correction, **c** with the "average" correction and **d** with the "lumi-average" correction. Only cell families from EBA are used

D.5   Other Sources of Systematic Uncertainty

The calibration transfer uncertainty is only one of the uncertainty sources that affect the luminosity determination. The other two main categories of systematic uncertainties in the ATLAS luminosity are:

vdM calibration   The uncertainty in the vdM calibration results from uncertainties in the beam population, beam conditions and from instrumental effects. In 2017 this uncertainty is 1.6%. Figure D.9a shows the scan-to-scan reproducibility, which belongs to the beam conditions category and is the largest uncertainty in the vdM calibration for 2017.

Long-term stability   The long-term stability and consistency over the year is computed by comparing LUCID with Tracking, TileCal D6 cells, FCal, and the LAr electromagnetic endcap (EMEC); it amounts to 1.3% for the 2017 preliminary luminosity estimate. This is shown in Fig. D.9b.

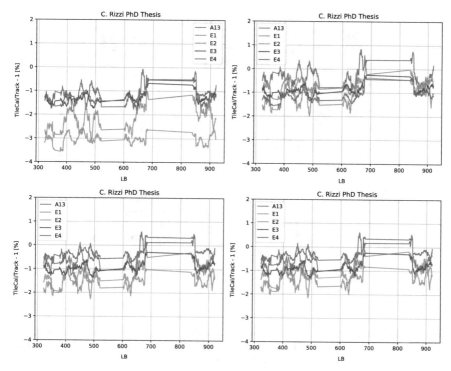

Fig. D.7 Fractional difference between the TileCal and Tracking luminosities for the vdM run **a** without any laser correction, **b** with the "jump" correction, **c** with the "average" correction and **d** with the "lumi-average" correction. Only cell families from EBC are used

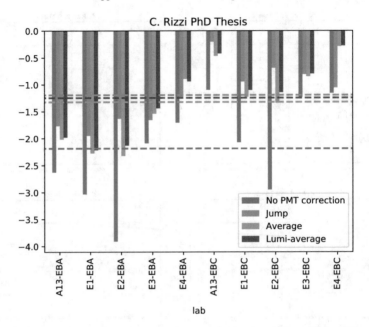

Fig. D.8 Relative difference in total integrated luminosity for the vdM run for the different types of PMT corrections. The dashed line shows the average of the difference over all the cell families considered, which is used to estimate the calibration transfer uncertainty

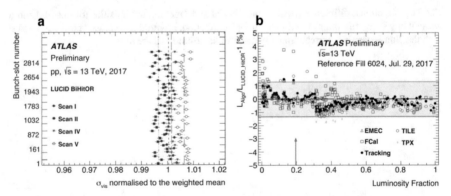

Fig. D.9 **a** Scan-to-scan reproducibility: ratios of bunch-by-bunch visible cross-sections to the weighted mean or all colliding bunch pairs and on-axis scans in the 2017 vdM runs. **b** Long-term stability: fractional differences in run-integrated luminosity between LUCID and the other measurements, plotted as a function of the cumulative delivered luminosity normalized to the 2017 total. The arrow indicates the run used to normalize the luminosity measurements from the other detectors to that of LUCID. Figures from Ref. [1]

D.6 Conclusion

The calibration transfer uncertainty is one of the major sources of uncertainty in the ATLAS luminosity measurement, and it has a relevant impact for analyses that rely on a precise luminosity measurement. The effect a non-linearity in the PMT response with the increase in luminosity in the calibration transfer uncertainty from TileCal has been studied. A correction has been derived that allows to reduce the calibration transfer uncertainty by 1% absolute. This correction has been applied to the computation of the luminosity uncertainty released in March 2018, leading to a calibration transfer uncertainty of 1.3%. This is summed in quadrature with the other uncertainty sources and the total luminosity uncertainty for the 2017 data-taking periods is estimated to be 2.4%.

References

1. ATLAS Collaboration (2019) Luminosity determination in pp collisions at $\sqrt{s} = 13$ TeV using the ATLAS detector at the LHC. ATLAS-CONF-2019-021. https://cds.cern.ch/record/2677054
2. ATLAS Tile Calorimeter system Collaboration, Abdallah J et al (2016) The Laser calibration of the Atlas Tile Calorimeter during the LHC run 1. JINST 11:T10005. https://doi.org/10.1088/1748-0221/11/10/T10005, arXiv:1608.02791 [physics.ins-det]
3. ATLAS Tile Calorimeter System Collaboration, Scuri F (2016) Performance of the ATLAS Tile LaserII calibration system. In: Proceedings, 2015 IEEE nuclear science symposium and medical imaging conference (NSS/MIC 2015): San Diego, California, United States. https://doi.org/10.1109/NSSMIC.2015.7581768
4. Gregorio GD (2017) Studies of the response stability for long term photomultiplier operation in the ATLAS hadronic calorimeter and a new method for photomultiplier gain measurements. Master's thesis, Universitá di Pisa, Italy

Printed in the United States
by Baker & Taylor Publisher Services